National
5

Chemistry

Stephen Jeffrey
John Anderson
Fran Macdonald
Barry McBride
Paul McCranor

HODDER
GIBSON
AN HACHETTE UK COMPANY

The Publishers would like to thank the following for permission to reproduce copyright material:

Photo credits

Fotolia, p. 1 (background) and Unit 1 running head image; Fotolia, figures 1.1, 1.2; Graham J Hills/SPL, figure 2.1; Fotolia, figure 2.2; Interfoto/Alamy, figure 2.3 (Democritus); Fotolia, figure 2.3 (John Dalton); Archive Pics/Alamy, figure 2.3 (J J Thompson); Image Asset Management Ltd/Alamy, figure 2.3 (Ernest Rutherford); Everett Collection Historical/Alamy, figure 2.3 (Niels Bohr); Pam McGookin, figure 3.20; Mark Boulton/Alamy, figures 5.1 a) and b); Pam McGookin, figures 5.4, 5.5; Fotolia, p. 45 (background) and Unit 2 running head image; Alex Segre/Alamy, figure 6.1; Fotolia, figure 6.2; Pam McGookin, figures 6.6, 6.7; Fotolia, figures 6.8, 6.9, 6.10; PAU, figure 6.14; Martin Shields/Alamy, figure 6.24; Pam McGookin, figure 6.25; James Holmes/Zedcor/SPL, figure 6.27; Fotolia, figure 6.29; Stockfolio/Alamy, figure 7.1; Fotolia, figure 7.2; John Anderson, figures 7.8, 7.9, 7.10; Inga spence/Alamy, figure 7.11; Tony Watson/Alamy, figure 7.12; John Anderson, figures 7.14, 7.18, 7.19; Fotolia, figures 7.20, 7.22, 7.23, 8.1; Jason Lindsey/Alamy, figure 8.2; Fotolia, figure 8.3; Whiteboxmedia limited/Alamy, figure 8.6; Pamela McGookin, figure 8.7; Elizabeth Fletcher, figure 8.9; Phil Degginger/Alamy, figure 8.10; Pam McGookin, figure 8.11; Fotolia, p. 75 (background) and Unit 3 running head image; Fotolia, figure 9.2; Qrt/Alamy, figure 9.5; Image Asset Management Limited/Alamy, figure 9.6; Fotolia, figures 9.8, 9.12, 10.1; Neil McAllister/Alamy, figure 10.2; Fotolia, figures 10.4, 10.12; PAU, figure 10.25; Fotolia, figure 10.26; Pam McGookin, figure 10.27; Fotolia, figures 11.1, 11.2; Gallo Images/Alamy, figure 11.4; Fotolia, figure 11.6; Nigel Cattlin/Alamy, figure 11.9; Moris Kushelevitch/Alamy, figure 11.10; Andy Buchanan/Alamy, figure 12.1; Photos 12/Alamy, figure 12.2; Fotolia, figure 12.3; TopFoto, figure 12.9; Archive Pics/Alamy, figure 13.1 (Orville Wright); NASA, figure 13.1 (Neil Armstrong); Empato/Alamy, figure 13.2; Paul Glendell/Alamy, figure 13.3.

Cartoons by Emma Golley, figures 3.3 (tug-of-war), 5.6, 7.13

Acknowledgements

Questions marked with an asterisk (*) are copyright Scottish Qualifications Authority and are reproduced with kind permission.

Every effort has been made to trace all copyright holders, but if any have been inadvertently overlooked the Publishers will be pleased to make the necessary arrangements at the first opportunity.

Whilst every effort has been made to check the instructions of practical work in this book, it is still the duty and legal obligation of schools to carry out their own risk assessments.

Although every effort has been made to ensure that website addresses are correct at time of going to press, Hodder Gibson cannot be held responsible for the content of any website mentioned in this book. It is sometimes possible to find a relocated web page by typing in the address of the home page for a website in the URL window of your browser.

Hachette UK's policy is to use papers that are natural, renewable and recyclable products and made from wood grown in sustainable forests. The logging and manufacturing processes are expected to conform to the environmental regulations of the country of origin.

Orders: please contact Bookpoint Ltd, 130 Park Drive, Abingdon, Oxon OX14 4SE. Telephone: (44) 01235 827720. Fax: (44) 01235 400454. Lines are open 9.00–5.00, Monday to Saturday, with a 24-hour message answering service. Visit our website at www.hoddereducation.co.uk. Hodder Gibson can be contacted direct on: Tel: 0141 848 1609; Fax: 0141 889 6315; email: hoddergibson@hodder.co.uk

© Stephen Jeffrey, John Anderson, Fran Macdonald, Barry McBride, Paul McCranor 2013

First published in 2013 by
Hodder Gibson, an imprint of Hodder Education,
An Hachette UK Company
2a Christie Street
Paisley PA1 1NB

Without Answers

| Impression number | 5 4 |
| Year | 2017 2016 2015 2014 |

ISBN: 978 1 4441 8430 3

With Answers

| Impression number | 5 4 |
| Year | 2017 2016 2015 2014 |

ISBN: 978 1 4441 8428 0

Cover photo: iStockphoto

Typeset in Minion Regular 11/14 by Integra Software Services Pvt. Ltd., Pondicherry, India

Printed in Dubai

A catalogue record for this title is available from the British Library

Contents

This book is designed to act as a valuable resource for students studying National 5 Chemistry. It provides a core text that adheres closely to the SQA specification. Each section of the book matches a unit of the specification; each chapter corresponds to a mandatory course key area. In addition to the core text, the book contains a variety of special features: *Activities*, *Questions* and *Checklists for Revision*. Questions marked with an asterisk (*) are from past SQA exam papers.

SQA assessment for National 5

Unit assessment

To pass each of the three course units you will have to:

- produce a scientific report of one of the experiments from the unit
- produce a short scientific report of a research investigation related to the unit
- answer a set of questions based on the key areas of the unit.

Course assessment and the assignment

At the end of the National 5 Chemistry course, you will sit a written test. The question paper will be worth a total of 80 marks (80% of the total mark) and it will have two sections:

- Section 1, titled 'Objective Test', will be worth 20 marks.
- Section 2, titled 'Paper 2', will contain restricted and extended response questions and will be worth 60 marks.

Open-ended questions

Paper 2 will normally contain two open-ended questions, each of which will be worth 3 marks. Open-ended questions can be used to assess whether or not you have truly grasped a chemical concept. In this type of question, you are required to draw on your understanding of key chemical principles to solve a problem or challenge. The open-ended nature of these questions is such that there is no unique correct answer. In addition to testing the extent of your chemical insight, these questions promote and reward creativity and analytical thinking.

In preparing for the course assessment, the following advice may be useful:

- In National 5 Chemistry examinations open-ended questions will *always* be identified by the wording *'Using your knowledge of chemistry, comment on...'*.
- Read the question carefully. Pay attention to diagrams, structural formulae and equations that have been included to help you answer the question.
- There will not be a single 'correct' answer. Markers will reward your understanding of chemistry.
- Reflect on the information provided in the question. Make sure that you answer exactly what the question is asking.
- Show your understanding of chemistry by drawing structural formulae, identifying functional groups, writing chemical equations or working out formulae.
- You may choose to present your answer as a paragraph, a set of bullet points or even as a diagram.
- If you have time at the end of the examination, check that what you have written answers the question asked.

Your response to an open-ended question will be assigned a mark according to how your understanding of chemistry matches the statements below:

- **0 marks:** The student has demonstrated **no understanding** of the chemistry involved.
- **1 mark:** The student has demonstrated a **limited understanding** of the chemistry involved.
- **2 marks:** The student has demonstrated a **reasonable understanding** of the chemistry involved.
- **3 marks:** The maximum available mark will be awarded to a student who has demonstrated a **good understanding** of the chemistry involved.

Practice open-ended questions

Following Unit 3 in this textbook, there are five practice open-ended questions for you to attempt.

The assignment

In addition to this, towards the end of your National 5 course you will be required to carry out an assignment. The purpose of the assignment is to allow you to carry out an in-depth study of a chemistry topic. You will choose the topic and investigate/research the underlying chemistry and the impact on society/the environment.

The assignment will be worth 20 marks (20% of the total mark). The majority of the marks (14 ± 2) will be awarded for applying scientific inquiry skills. The other marks (6 ± 2) will be awarded for applying related knowledge and understanding.

In preparation for this, you will carry out an investigation stage:

- select an appropriate chemistry topic within the set guidelines provided by SQA
- investigate/research the topic, focusing on applications and impact on society/the environment
- process the information/data collected.

You will then present evidence of:

- the process you have undertaken
- chemistry knowledge and understanding related to the topic investigated/researched
- the application of the topic
- a balanced evaluation of the impact on society/the environment
- a reasoned conclusion.

The assignment will be carried out under supervised conditions and there will be a time limit within which the investigation stage should be covered. Resources that you need will include access to sources of information relevant to your chosen topic. Your teacher/lecturer should discuss each stage of the assignment with you. There will also be a time limit within which you must complete the write-up. The assignment is an open-book task, so you will be permitted to use any resources from the investigation stage to complete the write-up.

Practice assignments

Following Unit 3 in this textbook, there are 14 practice assignments. These are mini-versions of the 'real thing' and so should *not* take the same time to complete! Your teacher/lecturer may allocate you one or more of these practice assignments as homework or as a self-study task. Whenever you are carrying out a practice assignment, remember to check your work against the above list and think honestly about the number of marks you would give yourself for the work you have done.

Each assignment brief will outline the task to be carried out. The stages to be covered in the assessment brief are: plan the investigation, identify resources, carry out research, select relevant information/data, gather relevant information/data and write up a report.

Your report should state the aim of the assignment, describe the process undertaken, present your data/results, include a *balanced* evaluation of the impact on society/the environment and a reasoned conclusion.

Unit 1

Chemical Changes and Structure

1 Rates of reaction

Imagine a world without chemistry. Your first thought might be that you would have a free period on your timetable at school, or that you would have picked a different subject instead. That probably is not all that appealing since you have chosen to study chemistry. Maybe you think it is a good thing though, that you would never need to wear safety goggles. Well, you would be right … sort of.

You would never need to wear safety goggles, because you would not be here. You would not exist. Nearly everything around you right now – your clothes, your bag, the products in your hair, the stool you are probably sitting on – would never have existed. The vast majority of the materials in our world were made by chemical reactions.

Chemical reactions are responsible for the world looking and progressing the way it does. Some reactions are over very quickly, like the burning of fuels or explosions (Figure 1.1), while others can take minutes, hours, days, months (Figure 1.2) or even millions of years.

The first part of this unit is about how chemists control the **rate** at which reactions take place. You may have learned about this in your lower school science lessons when you found out how to change the **speed** of a reaction, or in the National 4 Chemistry

Figure 1.2 Rusting is a slower reaction

course when you began to use the word 'rate' instead of 'speed'. In National 5 you will learn how to measure the rate of a reaction and express it as a number with a unit. This is a bit like saying a car is travelling at 70 miles per hour rather than simply that a car is travelling fast.

To change the rate of a reaction, in other words to speed it up or slow it down, it is necessary to change the reaction conditions. There are four variables that, when changed, will affect the rate of a chemical reaction:

- temperature
- concentration
- particle size
- use of a catalyst.

For any chemical reaction to take place, the reacting particles must **collide** with enough energy in order to combine and form new products. Changing one of the four variables above changes the way that these collisions take place, either by causing more collisions to occur, by increasing the energy of each collision, or both.

Temperature

If the temperature of a reaction mixture is increased, the particles have more energy and so move more quickly (Figure 1.3). Increasing the temperature increases the rate of reaction because the particles collide with more energy.

Figure 1.1 Explosions are fast reactions

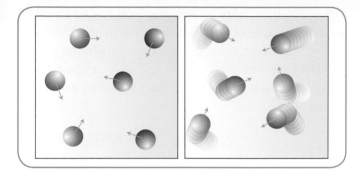

Figure 1.3 Particles at a higher temperature move more quickly so collide with greater energy

In summary:

- The higher the temperature, the faster the reaction.
- The lower the temperature, the slower the reaction.

Concentration

This variable can only be changed for reactants that are in solution. If the concentration of the reactants is increased, there are more reactant particles moving around in the same space (Figure 1.4). There will be more collisions and so the reaction rate is increased.

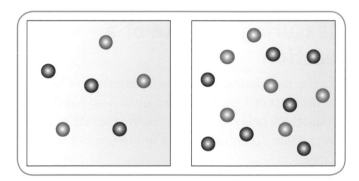

Figure 1.4 At a higher concentration, there are more collisions

In summary:

- The higher the concentration, the faster the reaction.
- The lower the concentration, the slower the reaction.

Particle size

This variable can only be changed for solid reactants. By crushing lumps into powder, for example, we decrease the particle size of a reactant and increase its surface area. The greater the surface area, the higher the chance

of collisions occurring and thus the faster the rate of reaction.

For example, think of a cube where the length of every side is 2 cm (Figure 1.5).

The area of one face of the cube will be $2 \times 2 = 4\,cm^2$.

Figure 1.5

The cube has six faces, so the total surface area is $4\,cm^2 \times 6 = 24\,cm^2$.

We could cut that cube horizontally and vertically along each face so that we have eight smaller cubes (Figure 1.6).

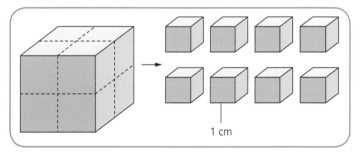

Figure 1.6

Each of the small cubes has a face area of $1\,cm \times 1\,cm = 1\,cm^2$.

The six faces give a total surface area for each smaller cube of $6\,cm^2$.

There are eight cubes so the total surface area is $6\,cm^2 \times 8 = 48\,cm^2$.

In summary:

- The smaller the particle size, the faster the reaction.
- The larger the particle size, the slower the reaction.

Catalysts

Catalysts change the rate of a reaction but are not used up in the reaction. Catalysts are used a lot in the chemical industry as they allow reactions to be carried out at lower temperatures, saving both energy and money. Platinum, rhodium and palladium are used as catalysts in car exhaust systems to convert harmful gases into less harmful gases.

Catalysts are also found in nature. Enzymes are biological catalysts and some of them are used within the body. For example, amylase in saliva helps to break down starch quickly to produce sugars which can give us energy. Enzymes can also be used in industrial processes such as the fermentation of sugars to produce alcohol.

Some catalysts work by allowing reactant molecules to 'adsorb' to active sites on their surface (see Figure 1.7), where they are held at an angle that allows the collision (and thus the reaction) to take place before the product molecules break off.

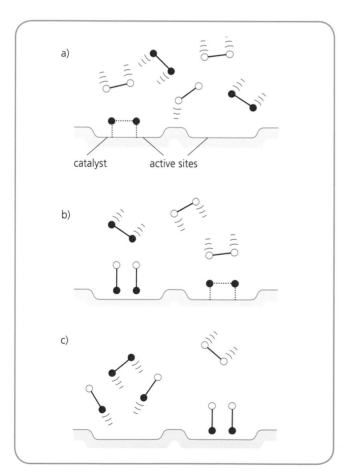

a)

catalyst   active sites

b)

c)

Other catalysts work by reacting with one or more of the reactants, meaning that temporarily the catalyst is 'lost'. However, once the reaction is over, the catalyst is regenerated, meaning that it can be reused. Imagine substances A and B react slowly to produce substance C:

$$A + B \longrightarrow C \qquad \text{slow reaction}$$

Catalyst Z is added to the mixture. Z reacts quickly with A to form a new substance AZ:

$$A + Z \longrightarrow AZ \quad \text{fast reaction (catalyst temporarily 'lost')}$$

Then AZ reacts in another fast reaction with substance B to form substance C and to reform substance Z:

$$AZ + B \longrightarrow C + Z \quad \text{fast reaction (catalyst regenerated)}$$

Combining the second and third equations gives:

$$A + B + Z \longrightarrow C + Z$$

So whether a catalyst just sits there allowing a reaction to take place more easily on its surface or actually gets involved in the process itself, one thing you can be sure of is that you will be able to use it again.

Measuring the rate of a reaction

The rate of a chemical reaction is a measure of how fast the reactants are being used up or how fast the products are being made. To measure the rate of any reaction accurately, we can follow one of three changes:

- changes in the mass of the reactants or products
- changes in the volume of the reactants or products
- changes in the concentration of the reactants or products.

Consider the reaction between chalk lumps and hydrochloric acid. This reaction produces carbon dioxide gas, which gives us two opportunities to monitor the course of the reaction and thus measure the rate of reaction.

Figure 1.7 a) adsorption, **b)** reaction, **c)** desorption

Method 1: Following the reaction by measuring the change in mass

The experiment should be set up as shown in Figure 1.8.

Figure 1.8 Measuring the rate of the reaction between chalk lumps and dilute hydrochloric acid

As the carbon dioxide gas is produced, it leaves the solution and so the total mass of the apparatus goes down. This loss in mass represents the mass of the carbon dioxide gas that has been produced and has escaped. If the initial mass of the apparatus was recorded and new mass readings were taken every 30 seconds, the results shown in Table 1.1 could be obtained.

Time (s)	Mass of apparatus (g)
0	198.1
30	196.7
60	195.8
90	195.4
120	195.0
150	194.7
180	194.4
210	194.1
240	193.9
270	193.8
300	193.8

Table 1.1

These results can be used to plot a graph (Figure 1.9) that shows how the rate of reaction changes with time. The graph can then be used to calculate the actual rate of the reaction. We can calculate the average rate of the reaction from start to finish, or we can calculate an average rate over any time period as the reaction is proceeding.

Figure 1.9

Method 2: Following a reaction by measuring the change in volume

The same experiment could be carried out using different apparatus that would allow us to measure the volume of carbon dioxide that is produced (Figures 1.10 and 1.11).

Figure 1.10 Using a gas syringe to monitor the volume of gas produced in a reaction

Figure 1.11 Collecting the gas produced in a reaction over water

Figure 1.12

If the volume of gas produced was noted every 30 seconds, the results in Table 1.2 could be obtained.

Time (s)	Volume of CO_2 produced (cm^3)
0	0
30	28
60	39
90	46
120	51
150	55
180	58
210	60
240	61
270	62
300	62

Table 1.2

A graph of these results would look like Figure 1.12. The section of the graph between 0 and 30 seconds has the highest gradient. The steepness of the line on the graph is an indication of the rate of reaction. This graph shows us that carbon dioxide is being produced very rapidly during the first 30 seconds.

Later in the reaction, say between 60 and 180 seconds, we can see that the line is less steep. Carbon dioxide

is still being produced, but not as quickly, and so the gradient is less than in the initial section of the graph.

Finally, between 270 and 300 seconds, the graph is horizontal – it has zero gradient. This means that the reaction has stopped and no more carbon dioxide is being produced.

The rate of the reaction is high to begin with because the concentration of the reactants is at its greatest; gradually, the rate slows down as the concentration of the reactants decreases.

Drawing graphs

Instead of an accurately plotted graph that can be used to calculate rate, sometimes you are provided with a sketch of a graph that shows the course of a chemical reaction. While these sketches do not always have numbers and units on their axes, they still provide information about the rate of the reaction as it proceeds.

Figure 1.13 shows two similar reactions between zinc metal and acid. This reaction produces hydrogen gas, the volume of which can be measured over time.

From the graph, we can see that at the start of the reaction, near the origin of the time axis, the red line (which represents zinc powder reacting with dilute

Figure 1.13

hydrochloric acid) indicates a faster reaction than the green line (the same reaction, but with zinc lumps used instead of powder) as the line is steeper. The end point of both reactions is clearly visible where the lines become horizontal, showing that no further hydrogen gas is being produced. The reaction involving zinc powder, as we would expect, finishes before that with the zinc lumps.

Calculating the average rate of a reaction

Over the course of a chemical reaction, the average rate can be calculated. As previously stated, we can calculate the average rate of the reaction from start to finish or we can calculate an average rate over any time period as the reaction is proceeding.

The average rate of a reaction is always given by the formula:

$$\text{average rate} = \frac{\text{change in measurable quantity}}{\text{change in time}}$$

The 'measurable quantity' we use can be the concentration, mass or volume of the reactants or products. As the rate of the reaction is constantly changing throughout the process, this calculation gives us the **average** rate over a specific period of time.

Worked example

Using the data from Table 1.1, calculate the average rate of the reaction between chalk lumps and acid between 30 and 120 seconds.

Reading from the table, at 30 seconds the mass of the apparatus was 196.7 g and at 120 seconds it was 195.0 g.

$$\text{average rate} = \frac{\text{change in mass}}{\text{change in time}}$$

$$= \frac{(196.7 - 195.0)}{(120 - 30)}$$

$$= \frac{1.7}{90}$$

$$= 0.019 \, g \, s^{-1}$$

Significant figures, decimal places and rounding

On a basic calculator, dividing 0.17 by 90 will show as 0.01888889. The answer given as 0.019 has been rounded to 2 significant figures or 3 decimal places (dps). An answer of 0.02 (1 significant figure or 2 dps) would also be acceptable.

Be careful when rounding. For example, 1.249 rounded to 1 dp would be 1.2. You cannot round to 2 dps first (giving 1.25), and then further round to 1 dp (giving 1.3). An *initial* figure of 1.25 rounded to 1 dp is 1.3.

Throughout this book, no hard and fast rule has been applied – most numerical answers are given to 1 decimal place, but occasionally, 2, 3 and sometimes even 4 decimal places are used. When writing your final answer in an examination, you should indicate the extent of any rounding in brackets to be on the safe side. Generally, an answer given to 2 dps is acceptable.

It is important to note that the units for a rate calculation change depending on the quantity that was measured. In the above example, the mass was measured in grams and the time in seconds, so the average rate is given the unit grams per second ($g \, s^{-1}$). You may also see this written as g/s.

Worked example

The volume of oxygen gas produced when hydrogen peroxide decomposes was measured and the results were plotted to give the graph shown in Figure 1.14.

Calculate the average rate of reaction over the first 25 seconds of the reaction.

Figure 1.14

$$\text{average rate} = \frac{\text{change in volume}}{\text{change in time}}$$

$$= \frac{(50 - 0)}{(25 - 0)}$$

$$= \frac{50}{25}$$

$$= 2\,\text{cm}^3\,\text{s}^{-1}$$

The unit in this example is cubic centimetres per second ($\text{cm}^3\,\text{s}^{-1}$) as the volume of oxygen was measured in cubic centimetres and the time in seconds.

Checklist for Revision

I can calculate the average rate of a reaction from graphs or tables of data to show the changes in the rate of a reaction as it progresses.

End-of-chapter questions

1 The course of a chemical reaction was followed by measuring the mass lost by the apparatus over time. The mass decreased by 50 grams in 200 seconds. What was the average rate of reaction in $\text{g}\,\text{s}^{-1}$?

   A 0.25

   B 0.333

   C 1

   D 4

2 Line A in Figure 1.15 refers to the reaction between magnesium and 0.1 mol l^{-1} nitric acid. What change in the reaction conditions could result in the reaction represented by line B?

   A Use magnesium lumps instead of powder.

   B Heat the acid.

   C Add a catalyst.

   D Stir the reaction mixture.

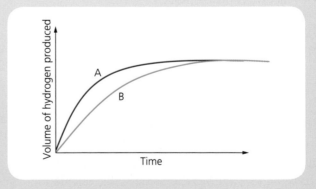

Figure 1.15

3 Hydrogen peroxide breaks down to form oxygen gas and water. You are asked to carry out this reaction and measure the reaction rate. Which combination of pieces of apparatus in Figure 1.16 would be *most* suitable to use in order to calculate the rate of the reaction?

syringe

Q

P

R

S

Figure 1.16

A P and Q
B S and Q
C P and R
D S and R

4 A student was following the rate of the reaction between chalk lumps and dilute acid by measuring the volume of carbon dioxide gas released over time. Which unit should she use when calculating the rate?

A $cm^3 s^{-1}$
B $g s^{-1}$
C $mol l^{-1} s^{-1}$
D $°C s^{-1}$

5 Figure 1.17 shows how the concentration of a reactant changes over time as products are being formed. What is the average rate of reaction between 20 seconds and 40 seconds in $mol l^{-1} s^{-1}$?

A 0.001875
B 0.0025
C 0.00375
D 0.005

Figure 1.17

6 Margarine is made in the chemical industry by the reaction between vegetable oil and hydrogen gas. To help the reaction to work effectively, the vegetable oil is heated and a powdered nickel catalyst is used. Suggest why the nickel catalyst is used as a powder rather than as a lump.

7 A student was investigating the rate of the reaction between copper(II) carbonate powder and dilute hydrochloric acid. He decided to measure the mass of carbon dioxide gas that was produced over time. He obtained the results given in Table 1.3.

Time (s)	Mass of carbon dioxide produced (g)
0	0
10	0.24
20	0.33
30	0.37
40	0.39
50	0.39

Table 1.3

a) Draw a diagram of the apparatus that he could have used to carry out the reaction.
b) Plot a line graph of the results shown above.
c) With reference to the graph, how could he tell that the reaction had finished?
d) Calculate the average rate of the reaction over the first 10 seconds.

8 Zinc lumps react with dilute hydrochloric acid in a test tube producing bubbles of hydrogen gas.
   a) What chemical test could be used to identify that the gas produced was hydrogen?
   b) Suggest two changes to the conditions that would result in a slower chemical reaction.
   c) Copy the graph in Figure 1.18 and add a line to represent the slower reaction.

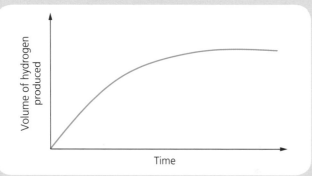

Figure 1.18

9 Figure 1.19 shows the change in the concentration of an acid as a chemical reaction occurs. Use the graph to calculate:
   a) the average rate of reaction in $mol\,l^{-1}\,min^{-1}$ over the first 2 minutes
   b) the average rate of reaction in $mol\,l^{-1}\,min^{-1}$ between 2 and 7 minutes.

Figure 1.19

10 A student was investigating the rate of reaction between hydrochloric acid and chalk. She weighed out 5 g of chalk lumps, placed them in $0.1\,mol\,l^{-1}$ hydrochloric acid and measured the volume of carbon dioxide produced at regular time intervals. After 180 seconds, the reaction was finished, and no chalk remained. Her results are shown in Table 1.4.

Time (s)	Volume of carbon dioxide produced (cm^3)
0	0
20	24
40	35
60	42
80	48
100	52
120	
140	57
160	59
180	60

Table 1.4

a) Plot a line graph of these results.
b) Use your graph to predict the volume of carbon dioxide produced after 120 seconds.
c) Add a dotted line to your graph to show what would happen if 5 g of chalk powder had been used instead of lumps.
d) Draw a labelled diagram of the apparatus the student could have used.

2   Atomic theory

Our world is mostly made up of **compounds** made from different combinations of **elements**. If we looked closely at an element through an extremely powerful microscope, such as a scanning electron microscope, we would see that an element is made of very tiny particles called **atoms**. Figure 2.1 shows gold atoms as seen through a scanning electron microscope.

In this part of Unit 1, we will look at atoms and what holds them together.

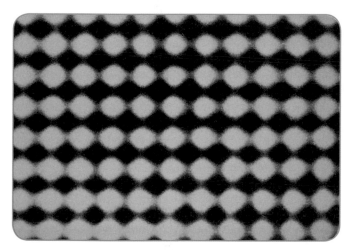

Figure 2.1 Gold atoms as seen through an electron microscope

The atom

Around 460BC Democritus, a Greek scientist, stated that all matter is made of atoms. In 1803 John Dalton (Figure 2.2), an English chemist, presented his atomic theory to the Royal Institution.

Dalton believed that atoms of different elements would have different masses. Based on this, he made some suggestions about atoms:

1   Elements are made from tiny particles called atoms.
2   All the atoms of one element are identical.
3   The atoms of one element are different from the atoms of any other element.
4   Atoms of one element can join with atoms of other elements to form compounds.

In the early twentieth century, however, scientists discovered that atoms are actually made up of even smaller sub-atomic particles called **protons**, **electrons** and **neutrons**.

Several scientists developed theories as to how these particles were arranged in the atom. In 1904 J. J. Thomson, who had discovered the electron in 1897, suggested that an atom was shaped like a sphere with the charged particles (electrons and protons) spread through it like the currants in a plum pudding, but this was proved to be incorrect. Another scientist, Sir Ernest Rutherford, suggested that the positive particles known as protons were clustered in the middle of the atom and he called this collection of particles the nucleus. This did not explain, however, why the mass of a single atom of most elements was much greater than the mass of the

Figure 2.2 John Dalton

protons in its nucleus and he suggested that there must be another particle that had yet to be discovered. The neutron was eventually discovered in 1932 by J. Chadwick. Neutrons are also found in the nucleus and they account for the extra mass. In 1913, a Danish scientist called Niels Bohr developed a theory of electron structure which described how the electrons are arranged around the nucleus in shells.

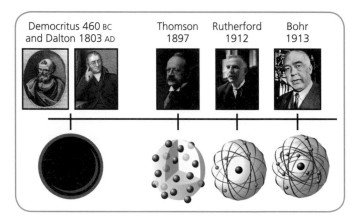

Figure 2.3 Timeline of the history of the atom

The work of these scientists has resulted in the atomic theory that we use today.

Atoms are made up of three smaller particles (Table 2.1):

Electrons: Negatively charged particles that spin around the positive centre of the atom in circles called **energy levels**. (Imagine how the Moon spins around the Earth.) Their mass is so small it is nearly zero.

Protons: Positively charged particles found in the nucleus of the atom (the centre). They are said to have a **relative** mass of 1.

Neutrons: Neutral particles (they have no electric charge) also found in the nucleus of the atom. They also have a relative mass of 1.

Particle	Relative mass	Charge	Location
electron	0	one negative (1−)	energy level
proton	1	one positive (1+)	nucleus
neutron	1	no charge	nucleus

Table 2.1 A summary of sub-atomic particles

The nucleus has an overall positive charge because it contains only protons (positive) and neutrons (neutral) (Figure 2.4).

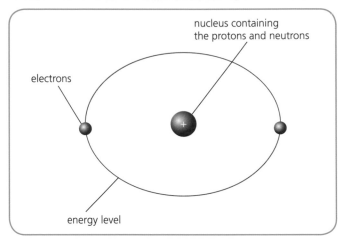

Figure 2.4 The basic structure of an atom

Atoms have no overall charge – they are neutral. This is because they contain equal numbers of positive protons and negative electrons. These opposite charges balance each other making the atom neutral.

The elements are arranged in the Periodic Table in order of increasing atomic number – hydrogen has the atomic number 1, helium 2, lithium 3 etc.

The atomic number of an element tells you how many protons there are in an atom of that element.

For example, a sodium atom has 11 protons (Figure 2.5).

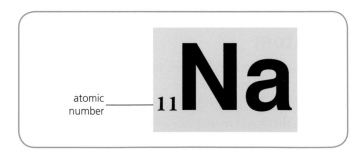

Figure 2.5 Sodium

Since atoms are neutral, this atom must also have 11 negatively charged electrons to balance the 11 positively charged protons. As you learned previously, electrons are found in energy levels around the nucleus of an atom. Each energy level can only hold a certain number of electrons. The first energy level (the one nearest the nucleus) can hold a maximum of two electrons with the others being able to hold up to a maximum of eight electrons. (This is only true for the first 20 elements.).

A sodium atom has 11 electrons, so two go into the first energy level (the one nearest the nucleus) and eight go into the second energy level, which leaves one that must go into the third or outer energy level (Figure 2.6). This information can be summarised as an **electron arrangement** 2,8,1.

The electron arrangements of other atoms can be found in the data booklet.

Elements in the same group of the Periodic Table (Figure 2.7) react in similar ways. This is because their atoms all have the same number of outer electrons. All alkali metals, for example, have one outer electron in their atoms (Figure 2.8). This makes them very reactive.

Group 1 – the alkali metals
Li   2,1
Na 2,8,1
K   2,8,8,1
Rb 2,8,18,8,1
Cs 2,8,18,18,8,1
Fr   2,8,18,32,18,8,1

Figure 2.8 Elements in the same group have the same number of outer electrons

Figure 2.6 How the electrons are arranged in a sodium atom

Figure 2.7 The Periodic Table

Activities

Draw diagrams similar to Figure 2.6 to show the electron arrangement of each of the first 20 elements. Use the data booklet to help you.

Now that we can use the atomic number to state the number of protons and electrons in an atom, we must also be able to work out how many neutrons there are in the nucleus of an atom. To do this we need the mass number of the atom.

The mass number of an atom is equal to the number of protons plus the number of neutrons.

13

The mass number is given at the top left of the element's symbol; for example, the sodium atom in Figure 2.9 has a mass number of 23.

Figure 2.9 Sodium

This way of displaying the mass number, atomic number and symbol for an atom is known as the **nuclide notation** and it can be used to calculate the numbers of protons, electrons and neutrons present. Figure 2.9 shows that the atomic number of sodium is 11. This tells us that a sodium atom has 11 protons and because its atoms are neutral, it also has 11 electrons.

The mass number of this sodium atom is 23.

Mass of protons = 11

Mass of electrons = 0 (electrons have almost zero mass)

Then the number of neutrons must be 12 to give sodium a mass of 23. (23 – 11 protons = 12 neutrons)

The number of each particle present can be worked out using your $P_{rotons} E_{lectrons} N_{eutrons}$!

$^{19}_{9}$F

P = 9
E = 9
N = 10 (19 – 9)

$^{27}_{13}$Al

P = 13
E = 13
N = 14 (27 – 13)

Some more examples are shown in Table 2.2.

Element	Atomic number	Mass number	Protons	Electrons	Neutrons
magnesium	12	24	12	12	12
potassium	19	39	19	19	20
carbon	6	12	6	6	6

Table 2.2

Isotopes

Think about the boys in your class. They may all be the same age, but they probably all have different masses. This is also true of atoms. Atoms of the same element can have slightly different masses.

Isotopes are atoms with the same atomic number but different mass numbers.

For example:

$^{12}_{6}$C and $^{13}_{6}$C

These two carbon atoms are **isotopes**; they have the same number of protons but different numbers of neutrons.

For carbon-12:	For carbon-13:
P = 6	**P** = 6
E = 6	**E** = 6
N = 6 (12 – 6)	**N** = 7 (13 – 6)

All elements have isotopes, so the mass of one atom of any particular element given in the data booklet is called the **relative atomic mass (RAM)**, which is the **average mass** of all the isotopes of a single element. The value given for the RAM in the data booklet is sometimes not a whole number because it is an average.

The relative atomic mass can be used to give an indication of the relative abundance of an element's isotopes.

Copper has two isotopes: copper-63 and copper-65. The RAM of copper is 63.5. This indicates that copper-63 is the more abundant isotope due to the fact that 63.5 is closer to 63 than to 65.

Bromine has two isotopes: bromine-79 and bromine-81. The RAM of bromine is 80. This indicates that there are equal amounts of both isotopes, because the average of 80 is exactly half way between the masses of the two isotopes.

Ions

Atoms are neutral because they have an equal number of positive protons and negative electrons. When there is an imbalance of electrons to protons, an **ion** is formed. Ions are charged particles and they are formed when atoms lose or gain electrons. Metal atoms lose electrons to form positive ions and non-metal atoms gain electrons to form negative ions (Table 2.3). Why should they do this?

The noble gases are a group of extremely unreactive elements. Apart from helium, which has atoms with only two electrons, which fill the first energy level, the electron arrangements for all the noble gases end in 8 – the outer energy level is full. This is said to be a **stable** electron arrangement and it is the reason why the noble gases are so unreactive. When atoms of other elements combine, their electron arrangements become like those of the noble gases – they achieve a full outer energy level. Neon atoms, for example, have eight outer electrons, whereas oxygen atoms only have six. To become stable like a noble gas, therefore, an oxygen atom must gain two electrons from another atom.

Metals	Non-metals
Magnesium	Chlorine
A magnesium atom has the electron arrangement of 2,8,2. Being a metal, it will lose two electrons to achieve the same electron arrangement as neon, 2,8. $$Mg \longrightarrow Mg^{2+} + 2e^-$$ This change creates an imbalance in the electron to proton ratio. For the ion $^{24}_{12}Mg^{2+}$: P = 12 E = 10 N = 12	A chlorine atom has the electron arrangement of 2,8,7. Being a non-metal, it will gain one electron to achieve the same electron arrangement as argon, 2,8,8. $$Cl + e^- \longrightarrow Cl^-$$ This change creates an imbalance in the electron to proton ratio. For the ion $^{35}_{17}Cl^-$: P = 17 E = 18 N = 18

Table 2.3

Checklist for Revision

- I can use nuclide notation for atoms and ions to calculate the numbers of protons, neutrons and electrons.
- I know what isotopes are and that the relative atomic mass (RAM) is the average mass of an atom of an element calculated from data about the element's isotopes and their proportions.

End-of-chapter questions

1 Isotopes of an element have a different number of
   A neutrons
   B protons
   C electrons
   D electrons and protons.

2 An atom of an element has the atomic number of 14 and a mass number of 28. The number of electrons in its atoms is
   A 7
   B 14
   C 28
   D 42

3 Which of the following electron arrangements is that of an element which has similar chemical properties to potassium?
   A 2,8,1
   B 2,8,2
   C 2,8,3
   D 2,8,4

4 Different atoms of the same element have identical
   A mass numbers
   B atomic numbers
   C numbers of neutrons
   D nuclei.

5 Which of the following numbers is the same for both lithium and fluorine atoms?
   A mass number
   B atomic number
   C number of outer electrons
   D number of occupied energy levels

6 The nuclide notation can be used to calculate the number of protons, neutrons and electrons that an ion contains. The nuclide notation of a silver ion is $^{107}_{47}Ag^+$.
   a) Copy and complete Table 2.4 to show the number of each particle that this ion contains.

Particles	Number
protons	
electrons	
neutrons	

Table 2.4

   b) A sample of silver is found to contain atoms of silver with different masses: ^{107}Ag and ^{109}Ag. What name is given to atoms of the same element with different masses?

   c) The relative atomic mass of silver is 108. What does this suggest about the relative abundance of these different atoms?

7 Table 2.5 shows two isotopes of chlorine.

Isotope	Number of protons	Number of electrons	Number of neutrons
$^{35}_{17}Cl$			
$^{37}_{17}Cl$			

Table 2.5

   a) Copy and complete the table for each of the isotopes.
   b) Chlorine has the relative atomic mass of 35.5. What does this suggest about the relative abundance of the isotopes?

8 Copy and complete Table 2.6.

Isotope	Mass number	Atomic number	Number of protons	Number of electrons	Number of neutrons
$^{31}_{15}P$					
		19			20
	15		7		

Table 2.6

9 The nuclide notation shows the atomic number and mass number of an isotope. The nuclide notation for an isotope of neon is: $^{21}_{10}Ne$.
   a) An isotope of calcium has the atomic number of 20 and a mass number of 42. Write the nuclide notation for this isotope of calcium.
   b) How many protons and neutrons does this isotope contain?
   c) All the isotopes of calcium are electrically neutral. What does this suggest about the proton to electron ratio of each isotope?

10 Copy and complete Table 2.7.

Ion	Number of protons	Number of electrons	Number of neutrons
$^{35}_{17}Cl^-$			
	8	10	8
$^{40}_{20}Ca^{2+}$			

Table 2.7

3   Bonding, structure and properties

Molecules

We will start this chapter by looking at the chemical bonds that hold atoms together in **molecules**. Molecules are usually made from non-metal atoms only. Water exists as molecules with the chemical formula H_2O. Each molecule contains two hydrogen atoms and one oxygen atom (Figure 3.1). The atoms in a molecule are held in place by **covalent bonds**.

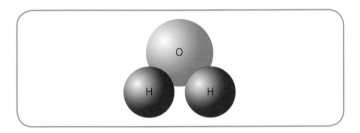

Figure 3.1 A simplified diagram of a water molecule

Covalent bonds

In Chapter 2 we looked at the noble gases and their full outer energy levels of electrons and discussed the idea that atoms of other elements achieve similar electron arrangements when they combine. When non-metal atoms combine with other non-metal atoms, the bonds formed are called covalent bonds.

A covalent bond is a shared pair of electrons between two non-metal atoms.

The atoms are held together because of the electrostatic force of attraction between the positive nuclei of each atom and the negatively charged electrons. This is better illustrated as a diagram (see Figure 3.2).

The saying 'opposites attract' can be used to describe how a covalent bond works. The positively charged nucleus of each atom is attracted to the negatively charged electrons. This creates a 'tug-of-war' effect. Both nuclei try to pull the

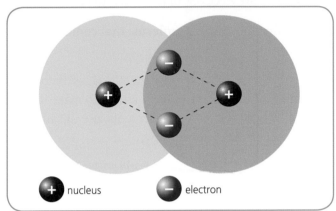

Figure 3.2 Covalent bonding in a molecule

electrons towards themselves, creating a strong bond that holds the atoms together. This is shown by the dashed lines in Figure 3.2.

Figure 3.3 represents a molecule of hydrogen. Hydrogen has the electron arrangement of 1. So if a hydrogen atom shares its electron with that of another atom of hydrogen, they both have two outer electrons, which results in them having the same electron arrangement as the noble gas helium.

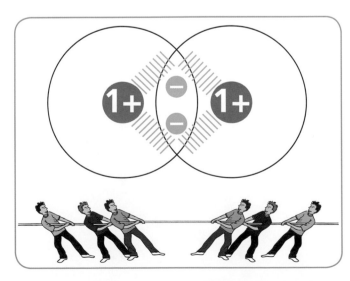

Figure 3.3 Covalent bonding can be likened to a tug of war

Remember the tug of war description; it will help you to remember how to describe exactly how a covalent bond holds atoms together.

Hydrogen is classed as a **diatomic element** as it exists not as single atoms of hydrogen, but as pairs of hydrogen atoms that share electrons in molecules to become stable.

A diatomic molecule is a molecule that contains only two atoms.

There are several other elements that exist as diatomic molecules, all of which you must know. To make it easy to remember all the diatomic elements, use this little mnemonic, or make up your own:

Fancy	Fluorine	F_2
Clancy	Chlorine	Cl_2
Owes	Oxygen	O_2
Him	Hydrogen	H_2
Nothing	Nitrogen	N_2
But	Bromine	Br_2
Ice	Iodine	I_2

For example, oxygen exists as diatomic molecules; that is why its formula is O_2. It is important to remember the names of all seven diatomic elements and to be able to draw a diagram of a molecule of each, showing all the outer electrons, as in the example drawn for hydrogen.

Figure 3.4 shows a molecule of fluorine. An atom of fluorine requires one electron to become stable, so it will share one electron with another atom of fluorine to form two stable atoms in a molecule.

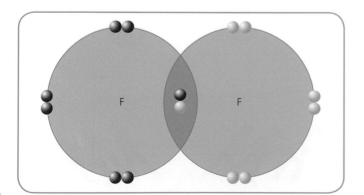

Figure 3.4 Bonding in a diatomic fluorine molecule

Oxygen atoms have the electron arrangement 2,6 and so require two electrons to become stable. Because of this, they must share two electrons with another oxygen atom (Figure 3.5). This means that oxygen forms a **double covalent bond**.

You may have noticed that so far we have only looked at how the bonds are formed in molecules of elements, but the same rules apply when dealing with the bonding in covalent compounds.

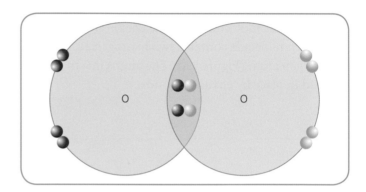

Figure 3.5 Bonding in a diatomic oxygen molecule

Methane (carbon hydride) has the chemical formula CH_4. This formula results from the atoms of carbon and hydrogen sharing electrons to become stable. An atom of carbon has four outer electrons (Figure 3.6) and therefore requires a further four to achieve a stable electron arrangement like a noble gas.

Figure 3.6 A carbon atom

A hydrogen atom has only one electron (Figure 3.7), so it requires only one more electron to achieve a stable electron arrangement.

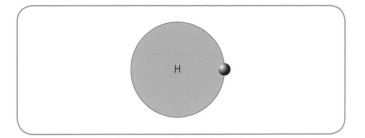

Figure 3.7 A hydrogen atom

For the two to combine and form a stable compound in which both hydrogen and carbon have stable electron arrangements, the carbon atom requires four hydrogen atoms to supply the four electrons (Figure 3.8).

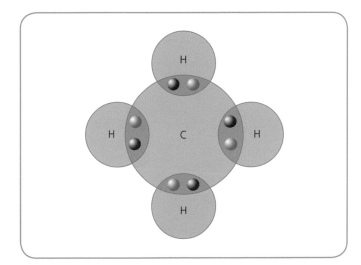

Figure 3.8 Bonding in a methane molecule

Sharing electrons in this way allows both hydrogen and carbon to have stable electron arrangements and this is the reason that methane (carbon hydride) has the formula CH_4.

Shapes of molecules

The repulsion between the pairs of electrons in the bonds results in covalent molecules having distinctive shapes. The bonds will move away from each other to create the shapes of the molecules that follow.

The three-dimensional shapes of molecules can be difficult to draw. To overcome this problem, chemists use the symbols shown in Figure 3.9 when drawing the shapes of molecules.

This represents a bond coming out of the page (towards you).

- - - - This represents a bond going into the page (away from you).

———— This represents a bond in the same plane as the page.

Figure 3.9 Drawing three-dimensional shapes

Tetrahedral molecules

Let's start with the example of methane (CH_4) (Figure 3.10). Many molecules that have a formula like methane, in which one atom of an element is joined to four of another, will have a **tetrahedral** shape. For example, CCl_4 (tetrachloromethane) will also be tetrahedral in shape.

Figure 3.10 The tetrahedral structure of methane

Pyramidal molecules

Ammonia (NH_3) is an example of a molecule with a **pyramidal** shape (Figure 3.11). Many other molecules that have a formula like ammonia, in which one atom of an element is joined to three of another, will have this shape. For example, PCl_3 (phosphorus trichloride) will also be pyramidal in shape.

Figure 3.11 The pyramidal structure of ammonia

Bent or V-shaped molecules

Water (H_2O) is an example of a molecule with a bent shape (Figure 3.12). In molecules containing three atoms, either the atoms form a straight line, in which case the shape is said to be **linear**, or there is a bend in the shape, like a boomerang.

Figure 3.12 The bent structure of water

Linear molecules

Hydrogen fluoride (HF) is an example of a linear molecule (Figure 3.13). All two-atom (diatomic) molecules must be linear.

Figure 3.13 The linear structure of hydrogen fluoride

Properties of covalent substances

Almost all covalent substances, both elements and compounds, are gases, liquids or soft solids with low melting points. Why should this be? Most covalent substances exist as small molecules containing a certain number of atoms held together by covalent bonds. These covalent bonds are very strong. Each molecule exists separately from all other molecules of the same substance. However, in addition to the covalent bonds *within* molecules (**intramolecular bonds**), which hold the atoms together, bonds also exist *between* molecules (**intermolecular bonds**). These intermolecular bonds are much weaker than covalent bonds. So when a covalent substance melts, it is only these weak forces of attraction that must be broken (Figure 3.14).

Figure 3.14 Discrete covalent substances melt at low temperatures because the weak bonds between the molecules are easy to break

Covalent substances like the ones described above are known as **discrete molecular** or **covalent molecular** substances. A few covalent substances have a very different structure, in which all the atoms in a sample are bonded to one another in one enormous molecule, made up of a large number of atoms that would be impossible to count. Since the bonds in these molecules are strong covalent bonds, melting would mean having to break these bonds. As a result, such substances have very high melting points and they are known as **covalent network** substances.

Diamond has a covalent network structure of carbon atoms (Figure 3.15). The many carbon atoms are held together by strong covalent bonds, meaning that in addition to having a very high melting point, diamond is also extremely hard. There are only three covalent network elements (boron, carbon and silicon) and two covalent network compounds (silicon dioxide and silicon carbide) that you will meet in this course.

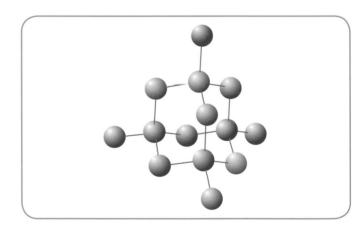

Figure 3.15 The covalent network structure of diamond

Discrete covalent substances generally have low melting points and boiling points. Like all covalent compounds, they can never conduct electricity, because molecules are not **charged** particles. Water is an exception to this, because a small number of water molecules split forming ions.

Ionic bonding

In Chapter 2, we looked at charged particles called ions and the fact that metals atoms form positive ions by losing electrons, and non-metal atoms form negative ions by gaining electrons. The attraction between oppositely charged particles is called an **electrostatic attraction**.

Ionic bonds are the electrostatic forces of attraction between positive ions and negative ions.

Ionic compounds

Ionic compounds have many different properties. For example, they all have very high melting points. Why should this be? Ionic compounds have what is known as a **lattice** structure. This is a regular arrangement of positive and negative ions.

Figure 3.16 shows a diagram of an ionic lattice. This ionic lattice has a cube-like structure. If you take a simple grain of salt (sodium chloride) and look at it under a microscope, you will see that the grain is a cube. It is not possible to count the actual numbers of sodium ions and chloride ions in the cube, but you can be sure that the numbers are the same – for every sodium ion there is a chloride ion. This is why the formula for sodium chloride is written as Na^+Cl^-. This represents the ratio of sodium ions to chloride ions in the lattice, in other words 1:1.

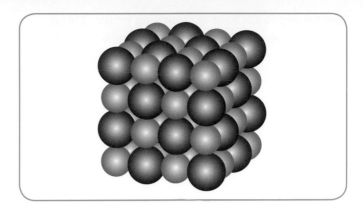

Figure 3.16 An ionic lattice structure

Many ionic compounds dissolve easily in water and when they do, their lattice breaks up completely to form free ions. The solutions formed will conduct electricity since ions are charged particles and in solution they are free to move. Moving electrical charges is an electric current. Melting ionic compounds also enables them to conduct electricity by freeing their ions to move.

Conduction can be tested in the laboratory as shown in Figures 3.17, 3.18 and 3.19.

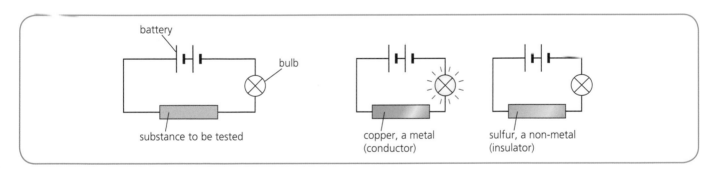

Figure 3.17 Testing a solid substance to see whether it conducts electricity

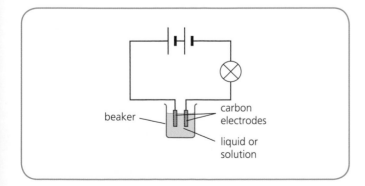

Figure 3.18 Testing a liquid or a solution to see whether it conducts electricity

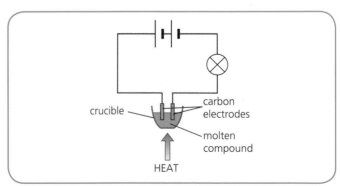

Figure 3.19 Testing a molten compound to see whether it conducts electricity

Ionic compounds have high melting points. This is because to melt an ionic compound, millions of strong ionic bonds must be broken, which requires lots of energy.

Not all ionic compounds form cube-like structures (Figure 3.20).

Figure 3.20 Crystals of various metal salts

Summary

Property	Ionic lattice	Covalent network	Covalent molecular
melting and boiling points	high	very high	low
state at room temperature	solid	solid	liquid, gas or low-melting-point solid
conduction of electricity	only when molten or in solution	never (except graphite)	never (although water is a poor conductor)

Table 3.1

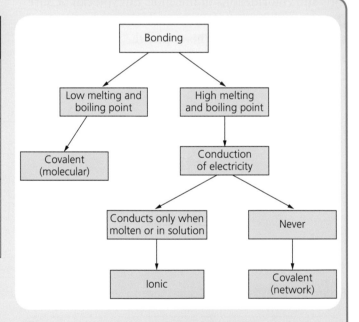

Figure 3.21 Flowchart to summarise types of bonding

Checklist for Revision

- I understand how covalent bonds are formed and the difference between covalent molecular substances and covalent networks.
- I understand how ionic bonds are formed and what ionic lattices are.
- I understand how the physical properties of substances are related to their bonding.

End-of-chapter questions

1 Which of the following substances exists as diatomic molecules?
   A Carbon tetrafluoride
   B Carbon dioxide
   C Calcium oxide
   D Carbon monoxide

2 Which of the following substances does *not* have a covalent network structure?
   A Diamond
   B Sulfur
   C Silicon dioxide
   D Silicon carbide

3 Which of the following has a covalent molecular structure?
   A Neon
   B Sulfur dioxide
   C Sodium chloride
   D Silicon dioxide

4 What type of structure is shown in Figure 3.22?

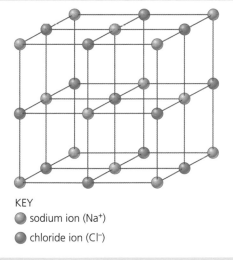

   KEY
   ● sodium ion (Na^+)
   ● chloride ion (Cl^-)

Figure 3.22

   A Covalent network
   B Covalent molecular
   C Ionic lattice
   D Tetrahedral

5 In a molecule of ammonia, all the atoms are held together by strong covalent bonds.
   a) Explain fully what a covalent bond is and how a covalent bond holds atoms together.
   b) Draw a diagram of an ammonia molecule showing all the electrons involved in holding the atoms together.
   c) Draw diagrams to show the shapes of methane, ammonia and water molecules.

6 The properties of three different substances are shown in Table 3.2.

Substance	Melting point (°C)	Boiling point (°C)	Conduction
A	−77	−33	no
B	1883	2503	no
C	773	1407	when molten

Table 3.2

   Copy and complete Table 3.3 using the letters to show the type of bonding present in each substance.

Substance	Bonding and structure
	ionic
	covalent network
	covalent molecular

Table 3.3

7 Although both contain covalent bonding, carbon dioxide is a gas at room temperature but silicon dioxide is a solid at room temperature. Explain why there is a large difference in melting points between the two covalent compounds.

8 Salt solution conducts electricity but glucose solution does not. What does this suggest about the types of bonding in salt and glucose? Explain your answer.

4 Chemical formulae

Chemical formulae are an easy way of passing on a lot of information about chemical compounds. Most non-chemists will recognise H_2O as the chemical formula for water, but how many understand what it is telling us about the compound? Few could tell us how to work out the formula for a compound. This section will give you all the tools that you need to tackle chemical formulae.

What does a formula tell us?

The formula of a covalent molecular compound gives the number of atoms present in a single molecule. For example, the formula for sulfur dioxide (SO_2) tells us that one molecule of the compound contains one sulfur atom and two oxygen atoms.

In ionic and covalent network compounds, the formula tells us something slightly different. The formulae for sodium chloride (NaCl) and silicon dioxide (SiO_2) give the simplest ratio of ions in the sodium chloride lattice (1:1) or atoms in the silicon dioxide network (1:2).

Formulae from names of compounds

Sometimes the name of the compound gives information about its chemical formula. The prefix given to either part of the compound's name gives the number of atoms of that element present in each molecule, as shown in Table 4.1.

Prefix	Number of atoms
mono-	one
di-	two
tri-	three
tetra-	four
penta-	five
hexa-	six

Table 4.1

Examples of chemical names that tell us their formula include carbon tetrachloride, which contains one carbon atom joined to four chlorine atoms, so it has the formula CCl_4. Similarly, water (H_2O) could also be known as dihydrogen monoxide!

> **Questions**
>
> 1. Write the chemical formulae for the following chemical compounds:
>   a) carbon monoxide
>   b) boron tribromide
>   c) phosphorus pentachloride
>   d) dinitrogen trioxide
>   e) xenon hexafluoride.

Valency

The **valency** of an element indicates its combining power, that is, the number of bonds that one of its atoms can form with others. As you have learned, atoms form chemical bonds to fill their outer energy levels of electrons, and so elements in the same group of the Periodic Table have the same valency (see Figure 4.1).

The noble gases have a valency of zero as they do not usually form bonds with other elements.

SVSDF system

The **SVSDF system** is a set of simple steps to follow which will help you to write the chemical formula for a compound. The letters stand for **S**ymbols, **V**alency, **S**wap, **D**ivide, **F**ormula.

Step 1: Write down the symbols of both the elements involved.
Step 2: Beneath each symbol, write its valency.
Step 3: Swap the valencies over.
Step 4: If the valencies can be simplified, divide them both by the smaller of the two numbers. If one of the numbers is already one, then they cannot be divided and simplified any further.
Step 5: Write the formula.

Group 1 Valency 1	Group 2 Valency 2	Group 3 Valency 3	Group 4 Valency 4	Group 5 Valency 3	Group 6 Valency 2	Group 7 Valency 1	Group 0 Valency 0
1 Hydrogen H 1							2 Helium He 2
3 Lithium Li 2,1	4 Beryllium Be 2,2	5 Boron B 2,3	6 Carbon C 2,4	7 Nitrogen N 2,5	8 Oxygen O 2,6	9 Fluorine F 2,7	10 Neon Ne 2,8
11 Sodium Na 2,8,1	12 Magnesium Mg 2,8,2	13 Aluminium Al 2,8,3	14 Silicon Si 2,8,4	15 Phosphorus P 2,8,5	16 Sulfur S 2,8,6	17 Chlorine Cl 2,8,7	18 Argon Ar 2,8,8
19 Potassium K 2,8,8,1	20 Calcium Ca 2,8,8,2	31 Gallium Ga 2,8,18,3	32 Germanium Ge 2,8,18,4	33 Arsenic As 2,8,18,5	34 Selenium Se 2,8,18,6	35 Bromine Br 2,8,18,7	36 Krypton Kr 2,8,18,8
37 Rubidium Rb 2,8,18,8,1	38 Strontium Sr 2,8,18,8,2	49 Indium In 2,8,18,18,3	50 Tin Sn 2,8,18,18,4	51 Antimony Sb 2,8,18,18,5	52 Tellurium Te 2,8,18,18,6	53 Iodine I 2,8,18,18,7	54 Xenon Xe 2,8,18,18,8
55 Caesium Cs 2,8,18,18,8,1	56 Barium Ba 2,8,18,18,8,2	81 Thallium Tl 2,8,18,32,18,3	82 Lead Pb 2,8,18,32,18,4	83 Bismuth Bi 2,8,18,32,18,5	84 Polonium Po 2,8,18,32,18,6	85 Astatine At 2,8,18,32,18,7	86 Radon Rn 2,8,18,32,18,8
87 Francium Fr 2,8,18,32,18,8,1	88 Radium Ra 2,8,18,32,18,8,2						

Figure 4.1 A section of the Periodic Table showing the valencies of the various groups

Worked examples

1 What is the formula for potassium oxide?

Symbols	K		O
Valency	1		2
Swap	2	✗	1
Divide	2		1
Formula		K_2O	

2 What is the formula for aluminium oxide?

Symbols	Al		O
Valency	3		2
Swap	2	✗	3
Divide	2		3
Formula		Al_2O_3	

No division has been necessary in these first two examples, because there is no common factor between the two numbers. In other words, they cannot be simplified further.

3 What is the formula for carbon sulfide?

Symbols	C		S
Valency	4		2
Swap	2	✗	4
Divide	$\dfrac{2}{2}$		$\dfrac{4}{2}$
	1		2
Formula		CS_2	

Remember that the valency of an atom is related to its group number in the Periodic Table.

2   Using the SVSDF method, write the chemical formulae for the following compounds:
   a) lithium sulfide
   b) potassium chloride
   c) calcium phosphide
   d) boron bromide
   e) magnesium nitride.

Formulae of compounds containing group ions

Group ions contain two or more atoms, and most of them have a negative charge. The formulae of some common group ions are shown in Table 4.2.

The number of charges on a group ion can be used as the valency of the ion to help with working out formulae.

Worked example

What is the formula for magnesium nitrate?
Since nitrate is in the 'one negative' column of the group ion table, it has a valency of one.

Symbols	Mg	NO_3
Valency	2	1
Swap	1	2
Divide	1	2
Formula		$Mg(NO_3)_2$

When a formula contains more than one group ion, the group ion *must* be written in brackets. In magnesium nitrate, the small two indicates that there are two nitrate ions (two NO_3^- groups) present in the formula.

With ionic compounds, we can take the formula one stage further and write ionic formulae. As you know, an ionic compound is made up of oppositely charged ions (usually a metal ion and a non-metal ion, or a metal ion and a group ion). The ionic formula includes the charges present on each of the ions. The charges are the same as the valency for the element or group ion, remembering of course that all metals will form positively charged ions and non-metals form negatively charged ions.

Worked example

Write the ionic formula of beryllium hydroxide.

Symbols	Be	OH
Valency	2	1
Swap	1	2
Divide	1	2
Formula		$Be(OH)_2$

Beryllium is a group two metal and has a valency of two. This means that a beryllium ion has a 2+ charge. The hydroxide ion is negatively charged and carries a 1– charge.

Ionic formula: $Be^{2+}(OH^-)_2$

One positive		One negative		Two negative		Three negative	
Ion	**Formula**	**Ion**	**Formula**	**Ion**	**Formula**	**Ion**	**Formula**
ammonium	NH_4^+	ethanoate	CH_3COO^-	carbonate	CO_3^{2-}	phosphate	PO_4^{3-}
		hydrogencarbonate	HCO_3^-	chromate	CrO_4^{2-}		
		hydrogensulfate	HSO_4^-	dichromate	$Cr_2O_7^{2-}$		
		hydrogensulfite	HSO_3^-	sulfate	SO_4^{2-}		
		hydroxide	OH^-	sulfite	SO_3^{2-}		
		nitrate	NO_3^-				
		permanganate	MnO_4^-				

Table 4.2

The other ionic compounds for which we have worked out formulae in this section are shown in Table 4.3, along with their names, formulae and ionic formulae.

Name	Formula	Ionic formula
potassium oxide	K_2O	$(K^+)_2O^{2-}$
aluminium oxide	Al_2O_3	$(Al^{3+})_2(O^{2-})_3$
magnesium nitrate	$Mg(NO_3)_2$	$Mg^{2+}(NO_3^-)_2$

Table 4.3

What is the formula for lead(IV) sulfide?

Symbols	Pb	S
Valency	4	2
Swap	2	4
Divide	$\frac{2}{2}$	$\frac{4}{2}$
	1	2
Formula		PbS_2

Questions

3 Using the formulae for group ions from Table 4.2, write the chemical formulae for the following compounds:
  a) magnesium hydroxide
  b) sodium hydrogencarbonate
  c) potassium permanganate
  d) aluminium sulfate
  e) ammonium phosphate.
4 Write the ionic formulae for each of the compounds in question 3.

Questions

5 Write the chemical formulae for the following chemical compounds:
  a) gold(III) oxide
  b) tin(IV) chloride
  c) iron(III) sulfide
  d) copper(I) bromide
  e) lead(II) iodide.

Formulae using Roman numerals

In some cases, particularly with transition metals, the valency of an element is not always the same when it appears in different compounds. This is indicated by showing the valency as a Roman numeral in brackets after the element's name (Table 4.4).

Roman numeral	Number
I	one
II	two
III	three
IV	four
V	five
VI	six

Table 4.4

Balanced equations

A **chemical equation** is an easy way of describing a chemical change, showing both the reactants and products. These equations are balanced when there is the same number of each type of atom on both sides of the equation.

As an example, consider what happens when calcium burns. Calcium atoms and oxygen molecules react together to make calcium oxide.

The word equation for the reaction would be:

$$\text{calcium} + \text{oxygen} \longrightarrow \text{calcium oxide}$$

The reactants, calcium and oxygen, are written to the left of the arrow and the product, calcium oxide, to the right.

If we simply replace each of the chemicals' names with their symbols or formulae, the equation will be unbalanced:

$$Ca + O_2 \longrightarrow CaO$$

Notice that there are unequal numbers of oxygen atoms (two) on the left-hand side compared with the right-hand side (one). To make things equal, you need to adjust the number of formula units of some of the substances until you get equal numbers of each type of atom on both sides. So:

$$Ca + O_2 \longrightarrow 2CaO$$

Now there are two oxygen atoms on each side of the equation. However, there are now two calcium atoms on the right-hand side, but only one on the left-hand side. So:

$$2Ca + O_2 \longrightarrow 2CaO$$

This is the balanced symbol equation.

Calcium oxide exists as an ionic lattice. Figure 4.2 shows a simplified version of its formation to help you understand the equation balancing process.

Figure 4.2

In the original equation, there were uneven numbers of oxygen atoms on the left- and right-hand side of the equation. Balancing the oxygen involves adding a two to the right-hand side. Doubling the number of units of calcium oxide produced means that we have to double the quantity of calcium on the reactant side of the equation too.

To balance equations, follow these simple rules:

Step 1: Check that all the formulae in the equation are correct.
Step 2: Deal with only one element at a time.
Step 3: Balancing involves writing numbers *in front of* formulae. You cannot change any of the smaller numbers *within* a chemical formula as this would change the formula.
Step 4: Check each element again; repeat step 3 if needed.

Worked example

Heptane (C_7H_{16}) burns in a plentiful supply of oxygen to produce carbon dioxide and water.

$$C_7H_{16} + O_2 \longrightarrow CO_2 + H_2O$$

This equation is unbalanced. There are seven carbon atoms on the left-hand side and only one on the right. To balance the carbon, write a '7' in front of the carbon dioxide.

$$C_7H_{16} + O_2 \longrightarrow \mathbf{7}CO_2 + H_2O$$

We now need to balance the hydrogen. We have 16 atoms of hydrogen on the left-hand side in C_7H_{16} and only two on the right. To balance the hydrogen atoms, write an '8' in front of H_2O.

$$C_7H_{16} + O_2 \longrightarrow 7CO_2 + \mathbf{8}H_2O$$

Nearly there! Now that the carbon and hydrogen atoms from heptane have been balanced, we are only left with oxygen still to be balanced. We have two atoms of oxygen on the left-hand side, but in total on the right-hand side we have 22 oxygen atoms (fourteen from $7CO_2$ and eight from $8H_2O$). This can be balanced by writing an '11' in front of the diatomic oxygen molecule on the left-hand side.

The balanced equation will be:

$$C_7H_{16} + 11O_2 \longrightarrow 7CO_2 + 8H_2O$$

Questions

6 Balance the following chemical equations:
 a) $C_2H_4 + O_2 \longrightarrow CO_2 + H_2O$

 b) $Ag_2S + Al \longrightarrow Al_2S_3 + Ag$

 c) $C_6H_{12}O_6 + O_2 \longrightarrow CO_2 + H_2O$

 d) $H_3PO_4 + Mg(OH)_2 \longrightarrow Mg_3(PO_4)_2 + H_2O$

 e) $CO + O_2 \longrightarrow CO_2$

The mole

Moles are small, furry mammals that live underground, famous for their burrowing skills and supposedly poor eyesight – but to a chemist, a **mole** is a unit of measurement.

When you try to balance equations, you possibly think in terms of counting the numbers of atoms of each element present on both sides. For example, when presented with:

$$Ca + HCl \longrightarrow CaCl_2 + H_2$$

you might say to yourself: 'there are two atoms of chlorine and two atoms of hydrogen on the right-hand side but only one of each on the left-hand side'. You may then write the balanced equation:

$$Ca + 2HCl \longrightarrow CaCl_2 + H_2$$

Once an equation has been balanced successfully, you can use it either to calculate the quantities of reactants needed for a reaction or the quantities of products produced. When it comes to doing the actual experiment, however, you do not count atoms. You could not start off with one atom of calcium in the above example! Enter the mole.

For most elements, 1 mole is the **relative atomic mass** (**RAM**) in grams, for example 12 g of carbon or 40 g of calcium. However, for the seven diatomic elements, 1 mole is *twice* the RAM in grams, for example 2 g of hydrogen or 32 g of oxygen. For compounds, 1 mole is the **relative formula mass** (**RFM**) in grams. RFM is the sum of the RAMs of each of the elements in the formula. Sometimes, you will see the term **gram formula mass** (**GFM**) used to represent 1 mole. This can refer to either elements or compounds.

Figure 4.3 Is this what chemists mean by a mole?

Questions

7 Calculate the mass of 1 mole of each of the following compounds:
   a) sulfur dioxide
   b) carbon hydride
   c) potassium permanganate
   d) calcium carbonate
   e) magnesium nitrate.

Worked examples

1 What is the mass of 1 mole of calcium chloride?

To calculate this all you have to do is work out the formula of calcium chloride using the SVSDF method, then add together the relative atomic masses of each element.

S	Ca		Cl		$CaCl_2$
V	2		1		
S	1		2		$40\,g + (2 \times 35.5\,g)$
D	1		2		$= 40 + 71$
F		$CaCl_2$			$= 111\,g$

Note that since there are two chloride ions in the formula, we must take that into account when calculating the mass.

1 mole of calcium chloride has a mass of 111 g.

2 What is the mass of 1 mole of ammonium phosphate?

S	NH_4		PO_4		$(NH_4)_3PO_4$
V	1		3		
S	3		1		$(3 \times 14\,g) + (12 \times 1\,g)$ $+ 31\,g + (4 \times 16\,g)$
D	3		1		$= 42 + 12 + 31 + 64$
F		$(NH_4)_3PO_4$			$= 149\,g$

1 mole of ammonium phosphate has a mass of 149 g.

Mole calculations

Once you have mastered calculating the formula mass of any compound, you can calculate the mass of a precise number of moles using a formula triangle (Figure 4.4). This formula triangle can also be used to calculate the mass of a compound if you already know the number of moles of the compound and have calculated the formula mass.

Using the formula triangle is straightforward. Cover whichever of the three variables you are trying to calculate and whatever is left will show you the equation to use.

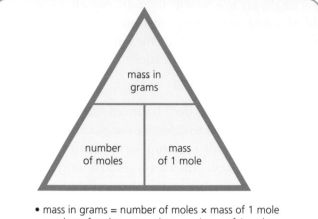

- mass in grams = number of moles × mass of 1 mole
- number of moles = mass in grams/mass of 1 mole
- mass of 1 mole = mass in grams/number of moles

Figure 4.4 The formula triangle

Worked examples

1 **How many moles are present in 25 g of calcium carbonate?**

The formula we need to use, given by covering up number of moles, is:

$$\text{number of moles} = \frac{\text{mass in grams}}{\text{mass of 1 mole}}$$

First, calculate the relative formula mass of calcium carbonate following the method used earlier.

The formula of calcium carbonate is $CaCO_3$.

$Ca = 40\,g$, $C = 12\,g$ and $O = (3 \times 16\,g)$

$RFM = 40 + 12 + 48 = 100$

mass (from the question) = 25 g

$$\text{number of moles} = \frac{\text{mass in grams}}{\text{mass of 1 mole}}$$

$$= \frac{25\,g}{100\,g}$$

$$= 0.25 \text{ moles}$$

2 **Calculate the mass of 2.5 moles of sodium nitrate (RFM = 85).**

Using the triangle to calculate mass gives the formula:

mass in grams = number of moles × mass of 1 mole

$$= 2.5 \times 85\,g$$

$$= 212.5\,g$$

Questions

8 Calculate the mass of the following:
   a) 3 moles of sodium oxide (Na_2O)
   b) 1.5 moles of potassium hydroxide (KOH)
   c) 0.25 moles of nitrogen dioxide (NO_2)
   d) 2.5 moles of sulfuric acid (H_2SO_4)
   e) 5 moles of magnesium chloride ($MgCl_2$).

9 How many moles are present in each of the following:
   a) 48 g of methane (CH_4)
   b) 150 g of calcium carbonate ($CaCO_3$)
   c) 6.4 g of sulfur dioxide (SO_2)
   d) 42.5 g of ammonia (NH_3)
   e) 146 g of hydrogen chloride (HCl)?

Calculations from balanced equations

A balanced chemical equation provides information about the reactants, how much you need to use (the number of moles), the products that will be made and how much you can expect to make (also in moles). For example, the equation:

$$Ca + 2HCl \longrightarrow CaCl_2 + H_2$$

tells us that 1 mole of calcium and 2 moles of hydrochloric acid can react to form 1 mole of calcium chloride and 1 mole of hydrogen. 1 mole of calcium weighs 40 g and 1 mole of calcium chloride weighs 111 g, so if you had 40 g of calcium, you could make 111 g of calcium chloride. But what if you only had 20 g of calcium – how much calcium chloride could you make? You can probably see that using only half as much calcium will lead to making half as much calcium chloride – 55.5 g. But what if you had, say, 8 g of calcium?

Questions

10 In each of the following problems, a balanced equation is provided. You will not always be so lucky!

a) Calculate the mass of hydrogen produced when 5 g of calcium reacts with an excess of dilute sulfuric acid.

$$Ca + H_2SO_4 \longrightarrow CaSO_4 + H_2$$

b) Calculate the mass of iron produced from 2 g of iron(III) oxide in the following reaction:

$$2Al + Fe_2O_3 \longrightarrow 2Fe + Al_2O_3$$

c) Calculate the mass of carbon dioxide formed when 0.8 g of methane is completely burned.

$$CH_4 + 2O_2 \longrightarrow CO_2 + 2H_2O$$

d) Calculate the mass of water produced by burning 10 g of hydrogen.

$$2H_2 + O_2 \longrightarrow 2H_2O$$

Step 1: Write a balanced chemical equation, unless already given.

$$Ca + 2HCl \longrightarrow CaCl_2 + H_2$$

Step 2: Identify the two chemicals referred to in the question and write the mole ratio

1 mole $\longrightarrow$ 1 mole

Step 3: Calculate the number of moles of the substance you have been given a mass for (Ca)

$$\text{number of moles} = \frac{\text{mass in grams}}{\text{mass of 1 mole}}$$

$$= \frac{8\,g}{40\,g}$$

$$= 0.2 \text{ mole}$$

Step 4: Use the mole ratio to calculate the number of moles of the substance you are trying to find (CaCl$_2$)

1 mole $\longrightarrow$ 1 mole

0.2 mole $\longrightarrow$ 0.2 mole

Step 5: Calculate the mass

mass in grams = number of moles $\times$ mass of 1 mole

$$= 0.2 \times 111\,g$$

$$= 22.2\,g$$

Mole calculations involving solutions

Many chemical reactions are not carried out using solids or liquids, quantities of which can easily be weighed. Some are carried out using solutions. You already know that the concentration of a solution is measured in moles per litre (mol l^{-1}). Now that you know what a mole is, perhaps this unit for concentration makes more sense. If you have 1 litre of a solution and it contains 1 mole of solute, then the concentration of the solution is obviously 1 mole per litre. Imagine you took a 500 cm^3 sample from this solution. What would the concentration of your sample be? The answer is still 1 mole per litre, because you will have 0.5 of a mole of solute in 0.5 of a litre of solution. The concentration of any solution can be calculated using the formula triangle given in Figure 4.5.

The volume used must be in litres. To convert cm^3 to litres, simply divide by 1000. For example, if the volume is given as 250 cm^3, then in litres this is 0.25 l (250/1000 = 0.25).

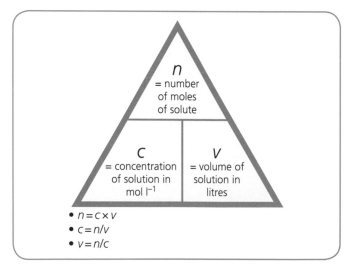

- $n = c \times v$
- $c = n/v$
- $v = n/c$

Figure 4.5 The formula triangle for a solution

Worked example

Calculate the concentration of a potassium nitrate solution if 3 moles are dissolved in 750 cm^3 of solution.

As before, covering concentration in our triangle gives the formula we want to use:

$$c = \frac{n}{v}$$

n = 3 moles

c = ?

v = 750 cm^3 = 0.75 l

$$c = \frac{n}{v}$$

$$= \frac{3}{0.75}$$

$$= 4 \text{ mol l}^{-1}$$

Questions

11 Calculate the concentration of the following solutions in mol l^{-1}:
   a) 0.25 moles of sodium hydroxide dissolved in 500 cm^3 of solution
   b) 3 moles of magnesium chloride dissolved in 2 litres of solution
   c) 1.5 moles of nitric acid dissolved in 1500 cm^3 of solution
   d) 6 moles of potassium bromide dissolved in 3 litres of solution
   e) 0.6 moles of potassium iodide dissolved in 250 cm^3 of solution.

More complicated examples

Sometimes you will meet a question that cannot be answered by either formula triangle individually. These problems must be solved using *both* of the triangles that we have learned about in one calculation.

Worked examples

1 What mass of calcium hydroxide is required to produce 500 cm³ of 0.1 mol l⁻¹ calcium hydroxide solution?

The question gives us a *concentration* and a *volume*, but these cannot be used alone to give us the required mass.

First, we must calculate the number of moles of calcium hydroxide that are required.

$n = ?$

$c = 0.1\,mol\,l^{-1}$

$v = 500\,cm^3 = 0.5\,l$

$n = c \times v$

$\quad = 0.1 \times 0.5$

$\quad = 0.05$ moles

The number of moles can be substituted into the other formula triangle and used to calculate the mass of calcium hydroxide required.

Remember that even if you are not given the formula mass, it can be calculated from the formula using the relative atomic masses of each element.

$Ca(OH)_2$

Ca = 40, O = (2 × 16), H = (2 × 1)

$\quad = 40 + 32 + 2$

$\quad = 74\,g$

mass in grams = number of moles × mass of 1 mole

$\quad\quad\quad = 0.05 \times 74\,g$

$\quad\quad\quad = 3.7\,g$

2 What is the concentration of a solution containing 5 g of sodium hydroxide in 250 cm³ of solution?

mass = 5 g

number of moles = ?

formula mass of NaOH = 23 + 16 + 1

$\quad\quad\quad\quad = 40$

$$\text{number of moles of NaOH} = \frac{\text{mass in grams}}{\text{mass of 1 mole}}$$

$$= \frac{5\,g}{40\,g}$$

$$= 0.125\ \text{moles}$$

Calculating the number of moles means that we can use $c = n/v$ to calculate the concentration.

$n = 0.125$ (calculated using the other triangle)

$c = ?$

$v = 250\,cm^3$ (given in the question) $= 0.25\,l$

$$\text{concentration of NaOH} = \frac{n}{v}$$

$$= \frac{0.125}{0.25}$$

$$= 0.5\,mol\,l^{-1}$$

Questions

12 Calculate the concentration of the following solutions in mol l⁻¹:
a) 10.1 g of potassium nitrate dissolved in 500 cm³ of solution
b) 5.55 g of calcium chloride dissolved in 50 cm³ of solution
c) 4 g of sodium hydroxide dissolved in 250 cm³ of solution
d) 69 g of potassium carbonate dissolved in 250 cm³ of solution
e) 1.49 g of ammonium phosphate dissolved in 20 cm³ of solution.

Checklist for Revision

● I can write chemical and ionic formulae, including formulae containing group ions.

● I can write balanced chemical equations.
● I can perform calculations based on balanced chemical equations.

5   Acids and bases

You will know about the **pH scale** either from lower school science lessons or from National 4 Chemistry. The pH of a substance is much more than a simple number. It gives important information about how the substance can be expected to react. In this topic, you will learn all about acids and their chemical opposites, bases. Many of the products you find in cupboards at home are acids or bases (Figure 5.1).

Figure 5.1 a) Household items containing acids

Figure 5.1 b) Household items containing bases

The pH scale

How acidic or basic a substance is (the pH of the substance) can be measured using the pH scale (Figure 5.2), a continuous range that stretches from below 0 to above 14. Most common pH values occur between 0 and 14.

- Solutions of acids have a pH of less than 7.
- Water and neutral solutions have a pH of exactly 7.
- Solutions of bases have a pH of more than 7.

The pH scale is actually a measure of the concentration of hydrogen ions in any given solution. The pH can be measured easily using a variety of pH indicators. Some, such as **universal indicator**, will show a colour for every pH value.

Other, more specialist, indicators are only used over a shorter range of pH and show fewer colour changes. These indicators are more useful for observing chemical reactions as they involve a sharper colour change. A summary of some common indicators is included in Table 5.1.

Indicator	Colours with pH changes
phenolphthalein	colourless below 8.3, pink above 10.0
methyl orange	red below 3.1, yellow above 4.4
thymolphthalein	colourless below 9.2, blue above 10.5
bromothymol blue	yellow below 6.0, blue above 7.6

Table 5.1

The pH of solutions

When we measure the concentration of hydrogen ions in a solution, we discover that acidic, alkaline (an alkali is a soluble base) and neutral solutions contain different concentrations of different ions. The important ions when dealing with pH are hydrogen ions (H^+) and hydroxide ions (OH^-).

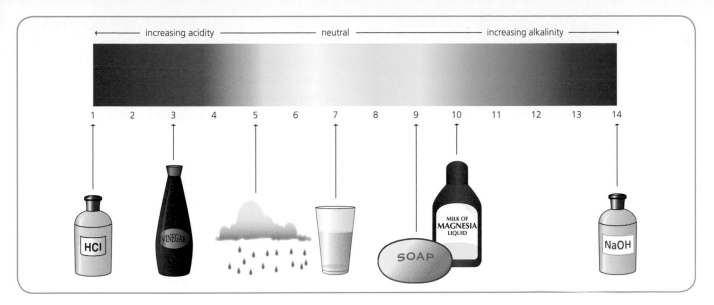

Figure 5.2 The pH scale

Water and neutral solutions

Water molecules can break down into hydrogen and hydroxide ions:

$$H_2O(l) \rightleftharpoons H^+(aq) + OH^-(aq)$$

This is a reversible reaction. A small proportion of water molecules break up to form hydrogen ions and hydroxide ions. Some of these hydrogen and hydroxide ions then react together again to form water molecules. A balance is established and the concentration of water molecules, hydrogen ions and hydroxide ions remains constant. It is the presence of this small number of ions that explains why water is able to conduct electricity.

In pure water and neutral solutions, the concentration of hydrogen ions is equal to the concentration of hydroxide ions.

Acidic solutions

Acidic solutions have a pH of less than 7. All acids are sources of hydrogen ions and contain H^+ ions in a greater concentration than OH^- ions. For any solution to be acidic, it must have a higher concentration of H^+ ions than OH^- ions.

The H^+ ions present in acidic solutions are responsible for acids reacting in the way they do. Acidic solutions conduct electricity well because they contain oppositely

charged ions that are free to move in solution and can carry charge.

Look at the formulae of the common laboratory acids in Table 5.2. When the ionic formulae are written, it is clear that they all contain H^+ ions.

Acid name	Formula	Ionic formula
hydrochloric acid	HCl	$H^+(aq) + Cl^-(aq)$
sulfuric acid	H_2SO_4	$2H^+(aq) + SO_4^{2-}(aq)$
nitric acid	HNO_3	$H^+(aq) + NO_3^-(aq)$
ethanoic acid	CH_3COOH	$H^+(aq) + CH_3COO^-(aq)$

Table 5.2 The formulae of some common laboratory acids

Strongly acidic solutions have a low pH. As acidity decreases, the pH number increases towards 7 on the pH scale.

Alkaline solutions

Alkalis are soluble bases. Alkaline solutions have a pH greater than 7. This is because all alkaline solutions contain an excess of hydroxide (OH^-) ions. Like acids, solutions of alkalis also conduct electricity because their ions are free to move in solution and can carry the charge.

The equation below shows what happens to sodium hydroxide when it is dissolved in water:

$$Na^+OH^-(s) \longrightarrow Na^+(aq) + OH^-(aq)$$

Metal hydroxides that are very soluble in water are all strong bases and produce solutions with a high pH when dissolved.

Table 5.3 shows the names and formulae of some common strong laboratory alkalis.

Alkali name	Ionic formula of solid	Ionic formula of solution
sodium hydroxide	Na^+OH^-	$Na^+(aq) + OH^-(aq)$
barium hydroxide	$Ba^{2+}(OH^-)_2$	$Ba^{2+}(aq) + 2OH^-(aq)$
lithium hydroxide	Li^+OH^-	$Li^+(aq) + OH^-(aq)$

Table 5.3 The formulae of some common laboratory alkalis

Dilution

Diluting an acid makes its pH go up towards 7 and diluting an alkali makes its pH go down towards 7. So adding water to an acid or alkali is one way of changing its pH. But what is actually happening to a solution when water is added? Why does the pH change at all? The answer is that although the number of moles (n) of hydrogen or hydroxide ions has not changed, the volume (v) in which they are dissolved has increased. As a result, the concentration of hydrogen or hydroxide ions has decreased and the pH will change accordingly.

Diluting an acid is not the most effective way of making an acid into a neutral solution, but let us look at how this could be achieved.

To change the pH of an acid by 1, a ten-fold dilution is required. If, for example, we begin with 1 cm³ of hydrochloric acid solution of concentration 0.1 mol l⁻¹ and measure its pH, we will find it has a pH of 1. To dilute the acid enough to change its pH to 2, we must add 9 cm³ of water to 1 cm³ of the acid to obtain a ten-fold dilution (1 cm³ + 9 cm³ = 10 cm³). To increase the pH of our 10 cm³ of acid to 3, we would need to add an additional 90 cm³ of water (Figure 5.4). Can you calculate the total volume of water that must be added to make our original 1 cm³ of acidic solution neutral? The data in Table 5.4 should help.

Volume	1 cm³	10 cm³	100 cm³				
pH	1	2	3	4	5	6	7

Table 5.4

Summary

All acidic solutions contain more hydrogen ions than hydroxide ions.

All alkaline solutions contain more hydroxide ions than hydrogen ions.

Neutral solutions contain an equal balance of hydrogen and hydroxide ions.

acidic solution   neutral solution   alkaline solution

Figure 5.3 The balance of ions in acidic solutions, neutral solutions and alkaline solutions

Figure 5.4 A 100-fold dilution causes a pH change of 2 units

Figure 5.5 Making an alkaline solution of magnesium oxide

When an acidic solution is diluted with water, the concentration of H^+ ions decreases and the pH of the solution increases towards 7. The solution is becoming less acidic. Similarly, when an alkali is diluted with water, the concentration of OH^- ions decreases. This causes the pH of the alkali to fall towards 7, making the solution less alkaline as more water is added.

Forming acids and bases

Acids and bases have a wide variety of uses, but how are they obtained?

An alkaline solution can be formed if a metal oxide is dissolved in water. Burning some magnesium ribbon in a gas jar of oxygen is an exciting experiment that produces a bright, white light (Figure 5.5a). If water and universal indicator are added and the gas jar shaken, the indicator shows us that the product of this combustion reaction dissolves in water to form an alkaline solution (Figure 5.5b).

This process can be represented as shown:

magnesium oxide + water $\longrightarrow$ magnesium hydroxide

$$MgO + H_2O \longrightarrow Mg(OH)_2$$

All other bases which are soluble in water (including the metal oxides of the alkali metals in Group 1 of the Periodic Table) will also dissolve to form solutions with a pH greater than 7. When they dissolve, the soluble metal oxides produce hydroxide ions, causing the pH to increase. Metal oxides outside of Groups 1 and 2 of the Periodic Table are insoluble in water.

Conversely, non-metal oxides, such as sulfur dioxide and nitrogen dioxide, will dissolve in water to form acidic solutions. These two oxides are products of the combustion of many fuels and they contribute to the creation of acid rain.

Summary

Soluble metal oxides produce alkaline solutions when dissolved in water.

Soluble non-metal oxides produce acidic solutions when dissolved in water.

Neutralisation

Neutralisation is the reaction of an acid with a base that results in the pH moving towards 7. It is a useful process that can be applied to situations in everyday life. The treatment of a bee sting (which contains methanoic acid) with baking soda (a common household base) is an example of a neutralisation reaction. Similarly, adding lime (a base) to lakes reduces their acidity and helps to counteract the effects of acid rain.

Figure 5.6 The pH lift

These reactions can occur between a wide variety of acids and bases, but no matter which acid you choose to neutralise which base (and vice versa), water is always produced as a result of these reactions.

Two general word equations that show the products of neutralisation reactions are given below. Remembering these will help you when it comes to working out what is being produced in neutralisation reactions.

$$\text{acid} + \text{metal oxide} \longrightarrow \text{salt} + \text{water}$$

$$\text{acid} + \text{alkali} \longrightarrow \text{salt} + \text{water}$$

Metal oxides and alkalis are two types of bases. Basic substances neutralise acids, resulting in the pH of the acid increasing towards 7 and water being produced. (Remember: a soluble base dissolves in water to form an alkaline solution.)

Naming salts

The salt formed depends on which acid and alkali have been used in the neutralisation reaction. The salt that you add to chips or cook with is called sodium chloride, but there are many other salts. Sodium chloride is readily mined or obtained from seawater but can also be made in the laboratory by a neutralisation reaction.

To name the salt that is produced by any neutralisation reaction, we must appreciate the chemical change that both the acid and the alkali undergo. To form the salt, the

metal ion from the alkali (or base) replaces the hydrogen ion from the acid (alkali to front, acid to back).

For example:

hydrochloric + sodium $\longrightarrow$ sodium + water
  acid   hydroxide   chloride

$$\text{HCl} + \text{NaOH} \longrightarrow \text{NaCl} + \text{H}_2\text{O}$$

The acid used will determine the type of salt that is produced. Different laboratory acids supply different negative ions to the process and so the name of the acid can be used to tell what the name of the salt will be, as shown in Table 5.5.

Name of acid	Salt name ends with ...
hydrochloric acid	... chloride
sulfuric acid	... sulfate
nitric acid	... nitrate

Table 5.5

The first part of the salt's name comes from the metal present in the base. Some examples are shown in Table 5.6.

Name of base	Salt name starts with ...
sodium hydroxide	sodium ...
potassium hydroxide	potassium ...
magnesium hydroxide	magnesium ...

Table 5.6

So, in a neutralisation reaction between sulfuric acid and potassium hydroxide, the products would be potassium sulfate and water. This can be represented by the following word equation:

sulfuric + potassium $\longrightarrow$ potassium + water
  acid   hydroxide   sulfate

$$\text{H}_2\text{SO}_4 + 2\text{KOH} \longrightarrow \text{K}_2\text{SO}_4 + 2\text{H}_2\text{O}$$

During neutralisation of an acid with an alkali, the H^+ ion from the acid joins with the OH^- ion from the alkali. This is why water is *always* formed in all such neutralisation reactions:

$$\text{H}^+ + \text{OH}^- \longrightarrow \text{H}_2\text{O}$$

Acids and metal carbonates

Acids can also be neutralised by metal carbonates. In these neutralisation reactions, there are three products. The hydrogen ions (H^+) from the acid react with carbonate ions (CO_3^{2-}) to form water and carbon dioxide gas. A salt is also produced. Your stomach contains hydrochloric acid with a pH of between 2 and 3. When you eat too quickly, this acid builds up, leading to acid indigestion. To relieve the discomfort, the extra acid must be neutralised. Common indigestion remedies contain bases such as sodium hydrogencarbonate or calcium carbonate. For example:

hydrochloric + calcium $\longrightarrow$ calcium + water + carbon
   acid   carbonate   chloride   dioxide

$$2HCl \quad + \quad CaCO_3 \longrightarrow \quad CaCl_2 + H_2O + \quad CO_2$$

The salt is named in the same way as before, taking the metal from the carbonate and the ending from the acid used.

In the experiment shown in Figure 5.7, we can prove that carbon dioxide is the gas produced by bubbling it through limewater. The limewater will change colour from colourless to cloudy white indicating the presence of carbon dioxide gas.

Figure 5.7 Limewater can be used to show that carbon dioxide is released when an acid reacts with a metal carbonate

Ionic equations

Observing the chemical change when an acid and base react together can be difficult to do. We get a more complete picture of the reaction that is taking place when we not only consider the symbol equation for the reaction, but also the **ionic equation**. Ionic equations are different from symbol equations because they show all the ions involved in the reaction separated out with their charges.

As an example, consider a solution of nitric acid being neutralised by lithium hydroxide.

$$HNO_3 + LiOH \longrightarrow LiNO_3 + H_2O$$

The salt formed is lithium nitrate. The ionic equation for the above process, showing all the charges involved, looks like this:

$$H^+(aq) + NO_3^-(aq) + Li^+(aq) + OH^-(aq) \longrightarrow$$
$$Li^+(aq) + NO_3^-(aq) + H_2O(l)$$

Inspection of this equation shows that lithium ions and nitrate ions appear on both sides – they have not changed in any way. They are referred to as **spectator ions**. Spectators who attend a football match or a concert do not actually participate. Spectator ions are ions that appear on both the left-hand side and the right-hand side of ionic equations. If we remove the spectator ions, the ionic equation above can be shortened:

$$H^+(aq) + \cancel{NO_3^-}(aq) + \cancel{Li^+}(aq) + OH^-(aq) \longrightarrow$$
$$\cancel{Li^+}(aq) + \cancel{NO_3^-}(aq) + H_2O(l)$$

Rewriting the equation without the spectator ions leaves us with the following chemical change:

$$H^+(aq) + OH^-(aq) \longrightarrow H_2O(l)$$

When an acid and metal carbonate react together to form a salt, water and carbon dioxide, the change occurring is slightly different. For example, when sulfuric acid is neutralised by sodium carbonate, the formulae equation is as follows:

$$H_2SO_4 + Na_2CO_3 \longrightarrow Na_2SO_4 + H_2O + CO_2$$

Rewriting this equation as an ionic equation gives:

$$2H^+(aq) + SO_4^{2-}(aq) + 2Na^+(aq) + CO_3^{2-}(aq) \longrightarrow$$
$$2Na^+(aq) + SO_4^{2-}(aq) + H_2O(l) + CO_2(g)$$

Note that the covalent substances, water and carbon dioxide, are included. Again, we can remove the spectator ions and rewrite the simplified equation:

$$2H^+(aq) + \cancel{SO_4^{2-}}(aq) + 2\cancel{Na^+}(aq) + CO_3^{2-}(aq) \longrightarrow$$
$$2\cancel{Na^+}(aq) + \cancel{SO_4^{2-}}(aq) + H_2O(l) + CO_2(g)$$

$$2H^+(aq) + CO_3^{2-}(aq) \longrightarrow H_2O(l) + CO_2(g)$$

Titrations

A **titration** is an experiment in which a *measured* volume of one solution is added to a *known* volume of another solution until the reaction between the two is complete. If the concentration of one of the solutions is known, then the concentration of the other can be calculated. In the National 5 course, we use titration to explore reactions between solutions of acids and alkalis.

When titrating, volumes of solutions are measured with great accuracy, using special pieces of glassware called **pipettes** and **burettes**. Unlike a measuring cylinder, a pipette can be used to measure only one precise volume of a solution. Pipettes come in various sizes, but they are all the same shape and each has only a single graduation mark.

In a titration, a pipette is normally used to measure an exact volume of alkali, which is then poured into a conical flask. Alkalis are seldom added to burettes because most alkalis are solutions of solids in water and if used in burettes, the tap can clog when the water evaporates if the user forgets to clean the burette thoroughly after use. Acid solutions are usually made by dissolving liquids and gases in water and so this problem does not arise.

Imagine you had a solution of an alkali of unknown concentration and you wanted to find out what the concentration was. Using a pipette (Figure 5.8), you would extract an exact known volume of the alkali and add it to a conical flask, followed by a few drops of a suitable indicator, chosen based on the predicted end-point of the titration. Some of the most common indicators are shown in Table 5.1 on page 34.

The acid solution that you are using to neutralise the alkali in the flask must be of known concentration. This solution is added to a burette. A typical school burette (Figure 5.9) has a scale marked in $0.1\,cm^3$ units, from $0.0\,cm^3$ at the top to $50.0\,cm^3$ at the bottom. This is the reverse of a measuring cylinder, where the scale goes upwards from the bottom! This is because ultimately an experimenter needs to know how much solution has been dispensed *from* the burette, rather than how much is still in it. If you fill one to the zero mark then open the tap to let the solution flow out, when you close the tap, the volume of solution dispensed is simply the new reading on the scale. However it is not necessary to fill burettes to the zero mark. As long as you note accurately, to one decimal place, what the reading on the scale is before you begin a titration, then note the new reading on the scale once the desired colour change has been observed showing that the titration has been completed, then the *difference* between these two readings will be equal to the volume of solution used. Exactly the same method can be used if you know the concentration of the alkali and want to find out the concentration of the acid.

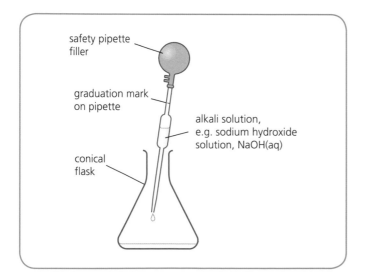

Figure 5.8 Pipetting an exact volume of alkali

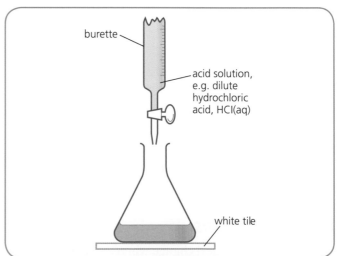

Figure 5.9 Running acid in from a burette

To ensure accuracy, a rough titration is usually carried out first. This provides an approximate end-point for the reaction, which can be accurately measured through repeatedly carrying out the titration until concordant results are obtained.

Titrating with accuracy

Here are a few key points to remember that will help you make accurate titrations:

- The conical flask should be sitting on a white tile. This will help you to observe colour changes more accurately.
- When emptying the pipette, the end should be tapped against the inner wall of the conical flask to ensure it empties completely.
- A rough titration should be carried out first to give an approximate end-point.

- After this, the first accurate titration can be carried out by adding solution from the burette to within 2 or 3 cm³ of the rough titration result, followed by drop by drop addition as the end-point is approached.
- All burette readings should be taken at eye level.
- The titration should be repeated until two concordant results (within 0.2 cm³ of each other) are obtained.
- When using the results for calculations, always *ignore* the rough titration result and calculate the average of the two concordant titrations.

Calculations based on titrations

There are several different methods that can be used to calculate an unknown concentration of an acid or alkali based on the results of a titration experiment. No matter which method you follow, the results of the calculation will be the same.

Worked example

A student is trying to calculate the concentration of a solution of sodium hydroxide. She titrates 20 cm³ portions of the alkali against a $0.1\,mol\,l^{-1}$ solution of sulfuric acid. Her results are shown in Figure 5.10.

Use the student's results to calculate the concentration of the sodium hydroxide.

A useful starting point, no matter which method you use for the calculation, is a balanced symbol equation for the process. The reaction that takes place is:

$$H_2SO_4 + 2NaOH \longrightarrow Na_2SO_4 + 2H_2O$$

The average titration of the student's experiment can be calculated as the average of her two concordant results. Her average titre is:

$$\frac{(15.0 + 15.0)}{2} = 15.0\,cm^3$$

Using the information known about the acid and the balanced equation, the unknown concentration can be calculated. This method uses the formula triangle that you learned earlier and the balanced equation.

Results

Titration	Start volume (cm³)	End volume (cm³)	Volume added (cm³)
rough	0.0	15.4	15.4
1st	15.4	30.4	15.0
2nd	30.4	45.4	15.0

Figure 5.10 The student's titration results

Balanced equation:

$$H_2SO_4 + 2NaOH \longrightarrow Na_2SO_4 + 2H_2O$$
$$\text{1 mole} \quad \text{2 moles}$$

The first job is to calculate the number of moles of acid that reacted:

number of moles $H_2SO_4 = c \times v$

$$= 0.1 \times 0.015$$

$$= 0.0015$$

From the equation: 1 mole of H_2SO_4 would react with 2 moles of NaOH. We have calculated that only 0.0015 moles of acid were present, therefore double that number (0.003 moles) of NaOH were neutralised.

Using the formula triangle again, we can calculate our final answer:

$$\text{concentration of NaOH} = \frac{n}{v}$$

$$= \frac{0.003}{0.02}$$

$$= 0.15 \, \text{mol} \, l^{-1}$$

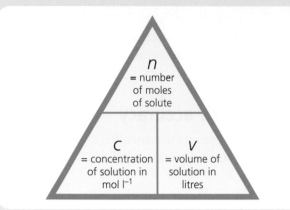

Figure 5.11 Formula triangle for a solution. Remember, this triangle gives:

- $n = c \times v$
- $c = n/v$
- $v = n/c$

Checklist for Revision

- I understand that water and solutions made using water as the solvent contain both hydrogen and hydroxide ions in varying concentrations.
- I understand that pH is related to the concentration of hydrogen ions in a solution.
- I understand that neutralisation is a reaction between an acid and a base which results in the pH moving to 7.
- I can calculate the concentration or volume of a solution of an acid or an alkali given the results of a titration.

End-of-chapter questions

1 Which of the following compounds could be produced in the reaction between an acid and an alkali?
   A Hydrogen chloride
   B Calcium hydroxide
   C Potassium sulfate
   D Nitrogen dioxide

2 Which of the following compounds would dissolve in water to produce a solution with a pH of less than 7?
   A MgO
   B $CaCO_3$
   C Na_2O
   D CO_2

3 What is the concentration of a solution containing 0.8 mol of potassium chloride in 200 cm³ of solution?
   A $0.004 \, \text{mol} \, l^{-1}$
   B $0.4 \, \text{mol} \, l^{-1}$
   C $4 \, \text{mol} \, l^{-1}$
   D $40 \, \text{mol} \, l^{-1}$

4 Identify the spectator ions in the following equation:
   $$2HCl + Mg(OH)_2 \longrightarrow MgCl_2 + 2H_2O$$
   A H^+ and Mg^{2+}
   B H^+ and OH^-
   C Mg^{2+} and Cl^-
   D Mg^{2+} and OH^-

5 A student titrated 25 cm^3 portions of 0.1 mol l^{-1} nitric acid solution against 0.1 mol l^{-1} sodium hydroxide. Her results are shown in Table 5.7.

Titration	Volume of sodium hydroxide (cm^3)
1	26.8
2	24.9
3	24.7

Table 5.7

a) Name a piece of apparatus she could use to measure 25 cm^3 portions of the nitric acid accurately.

b) What type of reaction is taking place between the nitric acid and the sodium hydroxide?

c) How many moles of nitric acid reacted?

d) What is the average volume of sodium hydroxide required to neutralise 25 cm^3 of the acid?

6 A student decided to produce potassium sulfate by first dissolving 276 g of potassium carbonate in water and then making the volume of the solution up to exactly 1 litre by adding more water. He added potassium carbonate solution to sulfuric acid to produce the desired salt.

a) Write a word equation for the reaction.

b) Rewrite this equation using chemical formulae.

c) What is the concentration of the potassium carbonate solution?

7 Explain the meaning of the following terms:

a) a base

b) an alkali.

8 A student wanted to find out the concentration of a solution of lithium hydroxide. She completed a titration experiment and found that 10.0 cm^3 of lithium hydroxide was neutralised by an average volume of 12.0 cm^3 of 0.1 mol l^{-1} hydrochloric acid solution. Calculate the concentration of her lithium hydroxide solution.

9 15.0 cm^3 of a solution of 0.1 mol l^{-1} NaOH neutralises 60.0 cm^3 of a solution of hydrochloric acid. What is the concentration of the acid?

10 25.0 cm^3 of a solution of 0.2 mol l^{-1} KOH neutralises 35.0 cm^3 of a solution of nitric acid. What is the concentration of the acid?

11 What volume of a solution of 0.1 mol l^{-1} NaOH is needed to neutralise 20.0 cm^3 of a solution of 0.3 mol l^{-1} hydrochloric acid?

12 What volume of a solution of 0.25 mol l^{-1} H$_2$SO$_4$ is needed to neutralise 40.0 cm^3 of a solution of 0.2 mol l^{-1} lithium hydroxide?

13 25.0 cm^3 of a solution of 0.2 mol l^{-1} Ba(OH)$_2$ neutralises 40.0 cm^3 of a solution of hydrochloric acid. What is the concentration of the acid?

14 25.0 cm^3 of a solution of 0.1 mol l^{-1} NaOH neutralises 60.0 cm^3 of a solution of hydrochloric acid. What is the concentration of the acid?

15 What volume of a solution of 0.01 mol l^{-1} potassium hydroxide solution is needed to neutralise 10.0 cm^3 of a solution of 0.15 mol l^{-1} sulfuric acid?

16 What volume of a solution of 0.005 mol l^{-1} HNO$_3$ is needed to neutralise 15.0 cm^3 of a solution of 0.1 mol l^{-1} lithium hydroxide?

This final question is very difficult – a real test of your knowledge!

17 75.0 cm^3 of a solution of 0.1 mol l^{-1} NaOH neutralises 25.0 cm^3 of a solution of an acid. The formula of the acid is H$_x$A and the concentration of the acid is 0.1 mol l^{-1}. What is the value of x?

Unit 2

Nature's Chemistry

6 Homologous series

It is hard to imagine the world without carbon compounds. From the fuels that power cars (Figure 6.1) to the deodorants and perfumes (Figure 6.2) that keep us smelling pleasant, carbon compounds are all around us.

Figure 6.1 Petrol and diesel contain carbon compounds such as alkanes and cycloalkanes

In this unit we will explore the structures and properties of four types of carbon compounds: hydrocarbons, alcohols, carboxylic acids and esters.

Figure 6.2 Many perfumes contain fruity smelling carbon compounds known as esters

To make our study of carbon compounds much easier, chemists have collected similar compounds into families. A family of compounds is known as a **homologous series**.

Alkanes

Hydrocarbons are compounds that contain hydrogen and carbon only. The **alkanes** are a homologous series of hydrocarbons where all the carbon-to-carbon bonds are *single* covalent bonds. The first member of the alkane homologous series is methane, the fuel in natural gas. Figure 6.3 gives a three-dimensional model of methane, showing its tetrahedral shape. The three-dimensional shape of methane is easier to picture if you build a molecule of methane from molecular models. This tetrahedral shape is difficult to draw by hand, so it is often more convenient to show methane by drawing its structural formula, as shown in Figure 6.4, or by writing its molecular formula, CH_4.

Figure 6.3 A molecule of methane

Figure 6.4 The full structural formula of methane

For large alkanes, such as hexane, we can also write a shortened structural formula. This is shown in Figure 6.5.

$$CH_3CH_2CH_2CH_2CH_2CH_3 \text{ or } CH_3(CH_2)_4CH_3$$

Figure 6.5 The full and shortened structural formulae of hexane

Table 6.1 lists the first eight members of the alkanes, showing their names and formulae.

Number of C atoms	Name	Molecular formula	Full structural formula	Shortened structural formula	State at room temperature (25 °C)
1	methane	CH_4		CH_4	gas
2	ethane	C_2H_6		CH_3CH_3	gas
3	propane	C_3H_8		$CH_3CH_2CH_3$	gas
4	butane	C_4H_{10}		$CH_3CH_2CH_2CH_3$	gas
5	pentane	C_5H_{12}		$CH_3CH_2CH_2CH_2CH_3$	liquid
6	hexane	C_6H_{14}		$CH_3CH_2CH_2CH_2CH_2CH_3$	liquid
7	heptane	C_7H_{16}		$CH_3CH_2CH_2CH_2CH_2CH_2CH_3$	liquid
8	octane	C_8H_{18}		$CH_3CH_2CH_2CH_2CH_2CH_2CH_2CH_3$	liquid

Table 6.1 The alkanes

47

General formula

For any homologous series we can write a **general formula**. This allows us to write the molecular formula for any member of a series where the number of carbon atoms is known. For the alkanes, the general formula is: C_nH_{2n+2}. For example, for an alkane with four carbon atoms ($n = 4$), the number of hydrogen atoms would be $(2 \times 4) + 2 = 10$. The formula would be C_4H_{10}, which is the correct formula for butane.

Physical properties

A **physical property** is something that can be observed or measured. Examples of physical properties include melting and boiling points, viscosity and colour. If you look up the melting and boiling point data for the alkanes, you will notice that there is a gradual increase from methane to octane. This is because bigger alkane molecules are more strongly attracted to each other than smaller alkane molecules and so it takes more energy to separate the bigger alkane molecules from each other. Hence, boiling points increase as the alkane molecules increase in size.

Chemical properties and uses

Chemical properties of substances refer to the chemical reactions which they take part in. The most common chemical reaction of the alkanes is combustion. You will be familiar with burning methane from experiments using the Bunsen burner (Figure 6.6) since methane is the fuel in natural gas. When alkanes burn completely, they produce carbon dioxide and water:

$$CH_4 + O_2 \longrightarrow CO_2 + H_2O$$

Complete combustion occurs in the Bunsen burner when the air hole is fully open. However, you will be aware that if you close the air hole, the flame colour changes from blue to yellow (Figure 6.7). Closing the air hole reduces the flow of oxygen. This causes incomplete combustion to occur, which accounts for the fact that the flame becomes sooty since carbon is one of the products of incomplete combustion. The toxic gas carbon monoxide is also produced during incomplete combustion.

The fact that alkanes burn makes them very useful as fuels. Table 6.2 lists some of the common uses for some of the alkanes.

Figure 6.6 Complete combustion of methane produces carbon dioxide and water

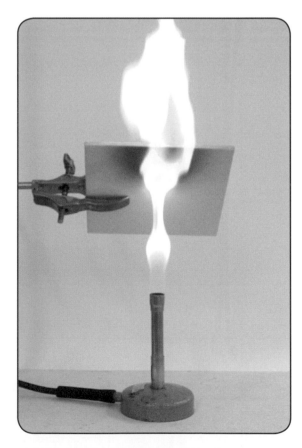

Figure 6.7 Carbon monoxide and carbon (soot) are also produced when methane burns in a poor supply of oxygen

Name of alkane	Use
methane	the fuel in natural gas for cooking and heating (Figure 6.8)
butane	the fuel used in barbecues and camping stoves (Figure 6.9)
octane	one of the compounds in petrol (Figure 6.10)

Table 6.2 Uses for some of the common alkanes

Figure 6.8 Methane is used at home for cooking on gas hobs

Figure 6.9 Butane or a butane/propane mix is used in camping stoves and can be bought in cylinders

Figure 6.10 Octane is one of the compounds found in petrol

Homologous series – a definition

Our study of the alkanes has revealed that the members of the alkane family share some common features:

- They have the same general formula, C_nH_{2n+2}.
- They have similar chemical properties; for example, they all burn.
- They show a gradual change in physical properties; for example, boiling points increase as the alkane molecules increase in size.

These features are key to being part of a homologous series. In other words, a group of compounds that have the same general formula, similar chemical properties and show a gradual change in physical properties can be said to belong to the same homologous series.

Isomers

If you use a set of molecular models to build the structure of a compound with molecular formula C_2H_6, there is only one possible structure. This is shown in Figure 6.11.

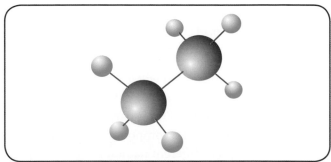

Figure 6.11 Molecular model structure of C_2H_6

49

However, if you try to build a molecular model structure of a compound with formula C_4H_{10}, you find that there are two possible structures, as shown in Figure 6.12.

Figure 6.12 Structures with the formula C_4H_{10}

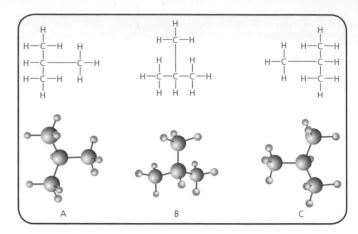

Figure 6.13 Rotating a structure – it is still the same isomer. Rotate molecule A 90° anticlockwise to obtain molecule B. Rotate molecule B 90° anticlockwise to obtain molecule C

Compounds that have the same molecular formula but different structural formulae are known as **isomers**. Isomers have different physical properties. For example, the boiling points of the two isomers of butane shown in Figure 6.12 are different. As the alkanes become larger in size, the number of isomers for each alkane increases too. For example, pentane has three isomers and hexane has five isomers.

Building molecular models is a useful way of determining how many isomers a compound has. For example, the second isomer of butane shown in Figure 6.12 can be drawn in many different ways, as shown in Figure 6.13, but they all show the exact same isomer! If you build the molecular model for this isomer, you will see that the structures simply show different rotations of the molecule. Try this to find out for yourself.

Systematic naming

If you look at the label of ingredients from a bottle of deodorant (Figure 6.14) or shower gel, you will notice that there are many exotic names.

In National 5 Chemistry you will learn how to identify and draw the structures of some of these compounds using **systematic naming**. This is particularly useful for identifying isomers of a compound as systematic naming gives isomers different names. For example, in Figure 6.12 only the top structure (the molecule

INGREDIENTS: Aqua, Aluminum Chlorohydrate, Steareth-2, Helianthus Annuus Seed Oil, Parfum, Steareth-20, Silica Dimethyl Silylate, Cholesterol, Lecithin, Citric Acid, Tocopherol, Potassium Lactate, Alpha-Isomethyl Ionone, Benzyl Alcohol, Benzyl Salicylate, Citronellol, Geraniol, Limonene, Linalool.

Figure 6.14 Deodorants contain carbon compounds such as alcohols and esters

with its four carbon atoms in a straight chain) is called butane. The lower branched molecule is called 2-methylpropane.

Here are some simple rules for naming branched alkanes:

1 Identify the longest chain of carbon atoms and name the alkane with this number of carbon atoms.
2 Identify the branch and name it according to the number of carbon atoms in the branch.
3 Number the branch so that it has the lower of two possible numbers.

Worked examples

1 Let us apply these rules to the isomer of pentane shown in Figure 6.15.

Figure 6.15 Naming an isomer of pentane

1 The longest chain of carbon atoms is four, so the alkane name is butane.
2 There is one branch. Branches are named after the alkane, as shown in Table 6.3. In this case, the branch is a methyl branch. If there had been two methyl branches, the prefix 'di' would have been used. In other words, it would be called 'dimethyl'.

Name of branch	Structure	Number of branches	Prefix
methyl	$-CH_3$	1	–
ethyl	$-C_2H_5$	2	di
propyl	$-C_3H_7$	3	tri

Table 6.3 Naming branches

3 There are two possible ways of numbering this structure, as shown in Figure 6.16. Applying rule 3, the lower of two numbers that we could give to the branch is 2.

The structure is called: 2-methylbutane.

Figure 6.16 The longest chain is numbered to give the branch the lower of two possible numbers

2 We will now apply the same rules to the structure given in Figure 6.17.

Figure 6.17

1 The longest chain has six carbon atoms: hexane.
2 There is one branch: ethyl.
3 The branch is on position 3.

This compound is called: 3-ethylhexane.

3 Applying the same rules to the structure given in Figure 6.18, we can work out that the compound is called 3,4-dimethylhexane.

Figure 6.18

Questions

1 Draw full structural formulae for the following compounds:
   a) 3-methylpentane
   b) 3,3-dimethylpentane
   c) 2,2,3-trimethylpentane.

Cycloalkanes

So far our study of hydrocarbons has dealt with the alkanes where each carbon is bonded to a neighbouring carbon atom in a straight chain. The **cycloalkanes** are a homologous series of hydrocarbons where the carbon atoms join together to form a closed ring. The first member of the cycloalkane series is cyclopropane, shown in Figure 6.19.

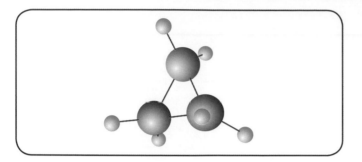

Figure 6.19 Cyclopropane, the first member of the cycloalkanes

You will see from Table 6.4 that the cycloalkanes are named after the parent alkane. For example, butane has four carbon atoms, so the cycloalkane with four carbon atoms is called cyclobutane.

General formula

The general formula for a cycloalkane is C_nH_{2n}.

Physical properties

As the cycloalkanes increase in size, their melting and boiling points increase.

Number of C atoms	Name	Molecular formula	Full structural formula	Shortened structural formula	State at room temperature (25 °C)
3	cyclopropane	C_3H_6			gas
4	cyclobutane	C_4H_8			gas
5	cyclopentane	C_5H_{10}			liquid
6	cyclohexane	C_6H_{12}			liquid
7	cycloheptane	C_7H_{14}			liquid
8	cyclooctane	C_8H_{16}			liquid

Table 6.4 The cycloalkanes

Chemical properties

Cycloalkanes burn to produce carbon dioxide and water.

Uses

Most cycloalkanes are used to make other compounds. For example, the main use of cyclohexane is for making nylon. Like the alkanes, cycloalkanes do not dissolve in water so they are good solvents for other compounds that do not dissolve in water.

The alkenes

The **alkenes** are a homologous series of hydrocarbons that contain at least one carbon-to-carbon **double** bond. Table 6.5 gives the names and formulae for the first seven alkenes.

General formula

The general formula for the alkenes is C_nH_{2n}.

Isomers and systematic naming

The fact that the cycloalkanes and the alkenes have the same general formula, C_nH_{2n}, allows us to conclude that cycloalkanes are isomers of the corresponding alkene with the same number of carbon atoms. For example, the cycloalkane with three carbon atoms is cyclopropane, C_3H_6. This is an isomer of propene, also C_3H_6. Both compounds have the same molecular

Number of C atoms	Name	Molecular formula	Full structural formula	Shortened structural formula	State at room temperature (25 °C)
2	ethene	C_2H_4		$CH_2{=}CH_2$	gas
3	propene	C_3H_6		$CH_2{=}CHCH_3$	gas
4	butene	C_4H_8		$CH_2{=}CHCH_2CH_3$	gas
5	pentene	C_5H_{10}		$CH_2{=}CHCH_2CH_2CH_3$	liquid
6	hexene	C_6H_{12}		$CH_2{=}CHCH_2CH_2CH_2CH_3$	liquid
7	heptene	C_7H_{14}		$CH_2{=}CHCH_2CH_2CH_2CH_2CH_3$	liquid
8	octene	C_8H_{16}		$CH_2{=}CHCH_2CH_2CH_2CH_2CH_2CH_3$	liquid

Table 6.5 The alkenes

formula but a different structural formula, as shown in Figure 6.20.

Figure 6.20 Two isomers with formula C_3H_6

Alkenes can also form isomers by moving the position of the double bond. For example, two isomers of butene can be formed by moving the double bond, as shown in Figure 6.21.

Figure 6.21 Two alkene isomers of butene

The isomers are named by numbering the carbon atoms from the end that gives the carbon of the double bond the lower number. For example, the isomer of pentene shown in Figure 6.22 would be called pent-2-ene rather than pent-3-ene.

Figure 6.22 Naming an isomer of pentene

Where there are branches, the double bond takes priority over the branch. This is shown in the examples in Table 6.6.

Physical properties

As the alkenes increase in size, their melting and boiling points increase.

Chemical properties

Like other hydrocarbons, alkenes burn to produce carbon dioxide and water. Unlike our previous two families, in

Compound structure	Name
	4-methylpent-2-ene
	2,4-dimethylpent-1-ene
	2,4,4-trimethylhex-1,5-diene

Table 6.6

which the carbon atoms were all bonded to one another by single covalent bonds, alkenes contain a carbon-to-carbon double bond. They are said to be **unsaturated**. The presence of the carbon-to-carbon double bond allows alkenes to take part in reactions known as **addition reactions**, where reacting molecules can 'add' across the reactive double bond, as shown in the examples which follow.

Adding hydrogen

Alkenes can react with hydrogen gas to produce alkanes (Figure 6.23).

Figure 6.23 Reaction of propene with hydrogen gas

This reaction is known as **hydrogenation**. The product alkane is said to be **saturated** because no more atoms can be added to the molecule. The alkene, on the other hand, is an example of an unsaturated molecule as more atoms can bond to it when the double bond breaks.

Hydrogenation is used in food manufacturing to change the texture of food, for example to change liquid oils into solids. This involves reacting **polyunsaturated** oils, which have several double bonds, with hydrogen to change some or all of the double bonds into single bonds. If you look at a cake or biscuit wrapper, you will often see the phrase 'contains hydrogenated vegetable oil' on the label of ingredients (Figure 6.24). However, due to health concerns, many food manufacturers have stopped using hydrogenated oils, which you will also see mentioned on the label.

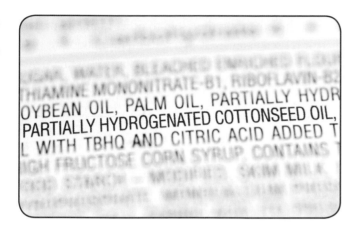

Figure 6.26 Reacting alkenes with bromine water

Uses

Making plastics

Alkenes can be made to react together to form very large compounds known as **polymers** (Figure 6.27). For example, ethene molecules can react with other ethene molecules to form poly(ethene); propene molecules can react with other propene molecules to form poly(propene). You will find out more about this reaction, **addition polymerisation**, and the uses of polymers in Unit 3.

Figure 6.24 Hydrogenation is used in the food industry

Adding bromine

A convenient laboratory test for the presence of double bonds in a molecule is to shake the compound with bromine water. Alkenes cause the bromine water to decolourise rapidly (Figure 6.25). With alkanes and cycloalkanes, the bromine water does not decolourise rapidly. The reaction of an alkene with bromine water is another example of an addition reaction: the bromine molecule adds across the double bond of the alkene, as shown in Figure 6.26.

Figure 6.27 Alkenes are reacted to form polymers

Figure 6.25 Alkenes rapidly decolourise bromine water

Industrial production of ethanol

In the next chapter, you will meet a new family of compounds known as the **alcohols**. The process of fermentation can be used to produce the alcohol **ethanol** from plants such as grapes, apples and barley. This is how the alcoholic drinks wine, cider and whisky are produced. Ethanol and other alcohols can also be produced by reacting alkenes with water. This addition reaction is known as **catalytic hydration**. For example, in order to meet the industrial demand for the alcohol ethanol, it is produced by the catalytic hydration of ethene (Figure 6.28):

$$C_2H_4 + H_2O \longrightarrow C_2H_5OH$$

Figure 6.28 Ethene reacts with water to produce ethanol

This involves passing ethene gas and steam over an acidic catalyst at high temperature and pressure. The INEOS chemical company is the largest producer of synthetic ethanol (ethanol produced by the catalytic hydration of ethene) in the world. Its site in Grangemouth, Scotland, produces over 300 000 tonnes of ethanol every year by this process. This ethanol is then sold to other manufacturers where it is used extensively as a solvent, for making medicines (Figure 6.29), printing inks and cosmetic products such as hairspray.

Figure 6.29 Ethanol is a useful solvent for making medicines and cosmetic products

Checklist for Revision

- I can name and write formulae for the alkanes, alkenes and cycloalkanes.
- I can use systematic naming to write a name for a structure.
- I can draw a structure for a compound from its systematic name.
- I know what is meant by the word 'isomer' and can identify isomers of a compound.
- I know what is meant by the term 'homologous series'.
- I know what is meant by the words 'saturated' and 'unsaturated'.
- I know some of the physical properties of alkanes, cycloalkanes and alkenes.
- I know that alkenes take part in addition reactions and I can draw the products formed when alkenes react with bromine, hydrogen or water.
- I can describe how bromine water can be used to distinguish between saturated and unsaturated compounds.

End-of-chapter questions

1 The grid below lists some common hydrocarbons.

A C_2H_4	B C_4H_{10}	C Hexane
D 2-methylhex-1-ene	E Cyclobutane	F Butene

  a) Identify the two alkanes.
  b) Identify the two saturated hydrocarbons with four carbon atoms.
  c) Identify the isomer of the compound in box F.
  d) Identify the three compounds that would rapidly decolourise bromine water.
  e) Identify the compound with seven carbon atoms.

2 Use the grid below to answer the following questions.

  a) Identify the two boxes that could represent butane.
  b) Identify the full structural formula for the compound formed when D reacts with hydrogen.
  c) Identify the fourth member of the alkenes.
  d) Identify the saturated isomer with formula C_4H_8.
  e) Identify the two compounds that could take part in addition reactions.

3 Draw full structural formulae for the second member of
  a) the alkanes
  b) the alkenes
  c) the cycloalkanes.

4 Two compounds, A and B, with formula C_3H_6 were reacted with bromine water. The bromine water decolourised rapidly when shaken with compound A, but did not decolourise rapidly with compound B. Draw full structural formulae for compounds A and B.

5 Draw full structural formulae for the following compounds:
  a) 2-methylbut-1-ene
  b) 3-methylpentane.

6 Write the systematic names for the following compounds.
  a)

Figure 6.30

  b)

Figure 6.31

  c)

Figure 6.32

7 **a)** State the general formula for the alkanes.
  b) Write molecular formulae for the alkanes with
   i) 12 carbon atoms
   ii) 80 hydrogen atoms.

8 The alkynes are a homologous series of hydrocarbons that contain a carbon-to-carbon triple bond. Table 6.7 lists the first three members of the alkynes.

Name	Structure
ethyne	H—C≡C—H
propyne	H—C≡C—C—H (with H above and H below the final C)
but-1-yne	H—C≡C—C—C—H (with H H above and H H below the final two C)

Table 6.7

  a) Write the molecular formula for each alkyne.
  b) The fourth member of the alkynes is pentyne. Draw a full structural formula for pent-1-yne.
  c) Write a general formula for the alkynes.

9* Alcohols can be oxidised by hot copper(II) oxide. The product is either an aldehyde or a ketone (Figure 6.33).

Alcohol	Structural formula	Type of product	Structural formula
ethanol	H—C—C—OH (H H above, H H below)	an aldehyde	H—C—C=O with H (aldehyde)
propan-1-ol	H—C—C—C—OH	an aldehyde	H—C—C—C=O with H
propan-2-ol	H—C—C—C—H (H OH H)	a ketone	H—C—C—C—H (H O H)
butan-2-ol	H—C—C—C—C—H (H OH H H)	a ketone	H—C—C—C—C—H (H O H H)

Figure 6.33

a) i) Aldehydes and ketones have the same general formula. Suggest a general formula for these compounds.
   ii) Write a general statement linking the type of product to the structure of the alcohol used.
b) Propan-1-ol can be formed by reacting an alkene with water. Draw the structure of the alkene used to produce propan-1-ol.
c) Draw a full structural formula for pentan-3-ol.
10* Dienes are a homologous series of hydrocarbons which contain two double bonds per molecule (Figure 6.34).

buta-1,3-diene   penta-1,3-diene

hexa-1,3-diene

Figure 6.34

a) What is meant by the term 'homologous series'?
b) Suggest a general formula for the dienes.
c) Write the molecular formula for the product of the complete reaction of penta-1,3-diene with bromine.
d) Draw a full structural formula for an isomer of buta-1,3-diene which contains only one double bond per molecule.

7 Everyday consumer products

Alcohols

Alcoholic drinks have been known of for thousands of years. The use and abuse of alcohol has been debated for much of that time and it can still be a great source of controversy. Many people use alcoholic drinks to celebrate social occasions (Figures 7.1 and 7.2) whilst other people do not drink alcohol. Unfortunately, most people in the UK are familiar with the tragic consequences of people consuming too much alcohol and the negative effects this can have on people and their families and friends. This drinking alcohol is known as **ethanol** and it is the first compound we will study in our review of the family of compounds known as the **alcohols**.

What is an alcohol?

The chemical structure of an alcohol can be obtained by replacing one of the hydrogen atoms of an alkane with an –OH group, the hydroxyl functional group, as shown in Figure 7.3.

Figure 7.3 Ethane and ethanol

As you can see from Table 7.1, replacing the final letter of the corresponding alkane by the name ending '-ol' gives the name of the alcohol.

Figure 7.1 Ethanol is the alcohol in alcoholic drinks

Figure 7.2 People celebrating with alcohol

Name	Structure	Shortened structural formula	Molecular formula
methanol		CH_3OH	CH_4O
ethanol		CH_3CH_2OH	C_2H_6O
propanol		$CH_3CH_2CH_2OH$	C_3H_8O
butanol		$CH_3CH_2CH_2CH_2OH$	$C_4H_{10}O$

Table 7.1 Names and formulae of common alcohols

59

General formula

The general formula for an alcohol is $C_nH_{2n+2}O$, although the shortened molecular formula $C_nH_{2n+1}OH$ is also used.

Isomers

For straight-chain alcohols, a simple way of creating isomers involves moving the –OH onto a different carbon atom. The alcohol is then named so that the –OH group has the lowest number. For example, the two structures shown for butanol would be named butan-1-ol (Figure 7.4) and butan-2-ol (Figure 7.5).

Figure 7.4 Butan-1-ol   **Figure 7.5** Butan-2-ol

Questions

1. Draw full structural formulae for the following alcohols:
   a) hexan-2-ol
   b) pentan-3-ol.
2. Explain why hexan-5-ol would be an incorrect name for an isomer of hexanol.
3. State the names of the following alcohols:

   a)

   Figure 7.6

   b)

   Figure 7.7

Uses of alcohols

Alcohols as solvents

One of the reasons we study alcohols in chemistry is because they are such useful compounds. The short-chain alcohols (methanol, ethanol, propanol and butanol)

are fantastic solvents as they can dissolve a variety of compounds. They are frequently used to dissolve compounds to make perfumes, deodorants, paints and dyes. As the alcohols evaporate easily, they are ideal liquids for the uses described. Once they have been applied to the skin, a whiteboard or a wall, the alcohol will evaporate to leave the solute. The short-chain alcohols are also frequently used in cleaning where their ability to dissolve a variety of compounds and evaporate easily makes them an ideal choice. For example, screen wipes, baby wipes, alcohol gels (Figure 7.8) and disinfectant wipes often contain propan-2-ol and ethanol.

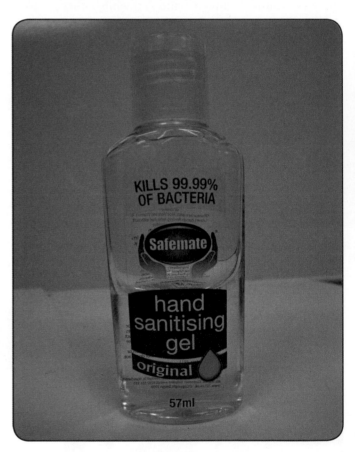

Figure 7.8 Handgels, deodorants and baby wipes contain alcohols

Alcohols as fuels

If you have ever witnessed an alcohol being set alight, you will be well aware of how flammable they are. A common trick used by magicians (and chemistry teachers!) is to set alight a banknote (Figure 7.9). Flames emerge from the banknote but it does not burn. This works because the solvent used is a mixture of ethanol and water. The water keeps the note cool while the ethanol burns. As long as the note stays cool, it will not burn. One of the problems

Figure 7.9 Setting alight a £10 note

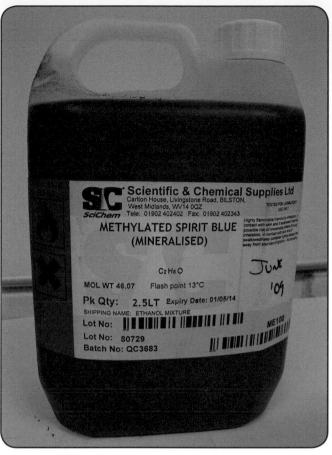

Figure 7.10 Methylated spirit contains ethanol and methanol and is commonly used as a fuel

with this demonstration is that the ethanol burns with a beautiful blue flame. In order to make the flames look more like 'fire', sodium chloride is often added since it burns with a yellow flame.

This demonstration gives us two important facts about burning alcohols:

- They burn easily.
- They burn with a clean blue flame.

The flame demonstrates that alcohols burn 'cleanly'. In other words, the alcohol reacts with oxygen to produce carbon dioxide and water in a complete combustion reaction. For example:

$$CH_3OH + 1\frac{1}{2}O_2 \longrightarrow CO_2 + 2H_2O$$

As alcohols burn to release a significant amount of energy, they are commonly used as fuels. For example, methylated spirits (Figure 7.10) is a fuel that contains ethanol and methanol. Methylated spirit camping stoves are commonly used because the fuel is cheap and readily available, and as they can burn most liquid fuels, it makes them an attractive purchase for explorers trekking through remote parts of the world. To stop people drinking the fuel, various additives are used including a dye and compounds that give the liquid a foul taste.

Ethanol as a fuel

One of the major disadvantages of fossil fuels, such as oil, coal and natural gas, is that they are not renewable. In contrast, ethanol is a renewable fuel since it can be produced by fermentation of sugar cane, which can be grown continuously. The ethanol produced can be used as a fuel or mixed with petrol or diesel to produce a fuel blend (Figure 7.11). For example, 'Gasohol' is a petrol containing between 10% and 20% ethanol.

The use of pure ethanol as a fuel has been encouraged in certain countries, such as Brazil, where it is economic to produce it by fermentation using surplus sucrose from sugar cane. Thus sugar cane can be regarded as a renewable source of engine fuel. Since the cane took in carbon dioxide during its growth, the combustion

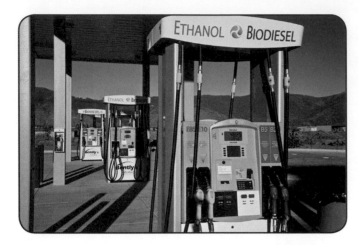

Figure 7.11 Petrol/ethanol blends and pure ethanol can be used as fuels for transport

of the ethanol to produce the same amount of carbon dioxide is not increasing the overall concentration of carbon dioxide in the atmosphere.

$$C_2H_5OH + 3O_2 \longrightarrow 2CO_2 + 3H_2O$$

Ethanol is therefore neutral in its contribution to 'greenhouse' gases.

Methanol as a fuel

Methanol is used as a fuel in racing cars in the USA (such as dragsters (Figure 7.12) and monster trucks) and has been suggested as a fuel for general use in cars too.

Figure 7.12 Methanol is used as a fuel in dragster racing cars

Carboxylic acids

Like many of the compounds we study in chemistry, the **carboxylic acids** are interesting because we find them in our everyday lives and make use of them to make a variety of products. For example, bee venom and ant stings contain methanoic acid (Figure 7.13), the first member of the carboxylic acids, whereas vinegar (Figure 7.14) contains a dilute solution of ethanoic acid, the second member of the series.

As you may have noticed the carboxylic acids we will study are named after the parent alkane with the term '-oic' added to the end of the name. Carboxylic acids contain the **carboxyl functional group**, shown in Figure 7.15.

Figure 7.13 Methanoic acid is found in bee venom and ant stings

Figure 7.14 Vinegar is a solution of ethanoic acid

Figure 7.15 Carboxylic acids contain the carboxyl functional group

The names and structures of some carboxylic acids are listed in Table 7.2.

Name	Structure	Shortened molecular formula	Molecular formula
methanoic acid	H—C—OH (with O double bonded)	HCOOH	CH_2O_2
ethanoic acid	H—C—C—OH (with H's and O double bonded)	CH_3COOH	$C_2H_4O_2$
propanoic acid	H—C—C—C—OH (with H's and O double bonded)	CH_3CH_2COOH	$C_3H_6O_2$

Table 7.2 Carboxylic acids

Questions

4 Draw full structural formulae for
   a) butanoic acid
   b) octanoic acid.
5 Name the following carboxylic acids:

a)

H—C—C—C—C—C—OH (with H's and O double bonded)

Figure 7.16

b)

H—C—C—C—C—C—C—OH (with H's and O double bonded)

Figure 7.17

6 State the general formula for the carboxylic acids.

Uses of carboxylic acids

Food preservation

As already mentioned, vinegar is a solution of ethanoic acid. It is widely used as a preservative in the food industry as the low pH of vinegar prevents bacteria and fungi growing which would spoil the food. Using vinegar in this way is an example of **pickling** (Figure 7.18). Vinegar is used to pickle eggs and various vegetables such as cucumbers.

Figure 7.18 Vinegar is used to pickle foods

Cleaning

As you discovered in the previous unit, acids such as hydrochloric and sulfuric are acidic because they produce hydrogen ions (H^+) when they dissolve in water.

$$HCl(aq) \longrightarrow H^+(aq) + Cl^-(aq)$$

Carboxylic acids also produce H^+ ions when added to water:

$$CH_3COOH(aq) \rightleftharpoons CH_3COO^-(aq) + H^+(aq)$$

Carboxylic acids are known as **weak acids** as they only partially ionise when they are added to water. Hydrochloric and sulfuric acid are known as **strong acids** since they break up completely into ions when they dissolve in water.

In areas of the UK where there is a high concentration of metal carbonates in the drinking water ('hard water'), metal carbonate deposits can build up in kettles, showerheads and taps. It has been found that vinegar is an ideal chemical for cleaning in such situations. Vinegar is highly effective at removing metal carbonate deposits because it contains ethanoic acid. Ethanoic acid can react with metal carbonates to produce soluble salts, water and carbon dioxide, which are washed away from the appliance being cleaned. The equation below shows the reaction between ethanoic acid and calcium carbonate:

$$2CH_3COOH(aq) + CaCO_3(s) \longrightarrow$$
$$(CH_3COO)_2Ca(aq) + H_2O(l) + CO_2(g)$$

Carboxylic acids in food

The food industry uses other carboxylic acids such as benzoic acid, C_6H_5COOH. This occurs naturally in cherry bark, raspberries and tea but it is synthesised for use as the food additive, E210. It acts as a preservative and is often used in fruit products and soft drinks. Sodium benzoate, E211, has similar properties and is used in bottled sauces and fruit juices.

Citric acid is another carboxylic acid commonly encountered when eating food. It is found in citrus fruits, such as lemons, and is responsible for the bitter taste you get when eating these fruits. An examination of soft drink labels usually reveals the presence of citric acid as its bitter taste is an essential component of many flavoured drinks. It is also widely used in cooking. For example, fruit salads are covered in lemon juice because the citric acid stops the enzymes working in bananas and apples, which would make the fruits turn brown on exposure to air.

Esters

One of the noticeable properties of many carboxylic acids is their unpleasant smell! For example, when butter becomes rancid (Figure 7.19), butanoic acid is formed which has a horrific smell. Other examples include 3-methyl-2-hexenoic acid which is found in human sweat and hexanoic acid which smells of goats. Yet, if these smelly acids are reacted with an alcohol, their smell can be changed completely into something quite pleasant. The fruity smelling compounds formed are known as **esters** and they are widely used in fragrances (Figure 7.20).

Figure 7.19 Butanoic acid is found in rancid butter

Figure 7.20 Many fragrances contain sweet-smelling esters

When ethanol reacts with ethanoic acid, the ester ethyl ethanoate and water are formed:

$$CH_3CH_2OH + HOOCCH_3 \rightleftharpoons CH_3CH_2OOCCH_3 + H_2O$$
ethanol   ethanoic   ethyl ethanoate
        acid

Forming an ester is an example of a **condensation reaction** since two molecules are reacted together to produce a larger molecule plus water. You will learn more about condensation reactions in Unit 3.

A variety of fruity smelling esters can be made by reacting some of the alcohols and carboxylic acids you are familiar with (Figures 7.21 and 7.22). For example, methanol and butanoic acid react to form methyl butanoate, which smells of apples. A pear smell can be made by reacting propanol and ethanoic acid together.

Apart from being used in perfumes, aftershave and deodorants, esters are mainly used as solvents. They

Figure 7.21 Formation of an ester

Figure 7.22 The smell of fruit can be made by reacting an alcohol with a carboxylic acid

make good solvents because they are able to dissolve compounds that are not soluble in water and they evaporate easily. For example, nail varnishes (Figure 7.23) often use ethyl ethanoate as the solvent. Once applied to the nail, the ester will evaporate to leave the varnish.

Figure 7.23 Esters are used as solvents

<hr>

Checklist for Revision

- I can recognise the hydroxyl group, –OH, in a structural formula and know that this indicates that the compound is an alcohol.
- I can name and draw straight-chain alcohols with up to eight carbon atoms.
- I know that alcohols are very good solvents.
- I know that alcohols can be used as fuels as they are highly flammable and burn with very clean flames.

- I can recognise the carboxyl group, –COOH, in a structural formula and know that this indicates that the compound is a carboxylic acid.
- I can name and draw straight-chain carboxylic acids with up to eight carbon atoms.
- I can state some uses of carboxylic acids.
- I know that alcohols and carboxylic acids can be used to form esters.
- I can state some uses of esters.

End-of-chapter questions

1 Which of the following shows the name of an alcohol with three carbon atoms?
   A Methanoic acid
   B Methanol
   C Propanol
   D Propanoic acid

2 Use the grid below to answer the questions that follow:

A Ethane
B Ethene
C Ethanol
D Ethanoic acid

   a) Which compound would have the lowest pH?
   b) Which two compounds could react together to form an ester?
   c) Which compound is found in vinegar?

3 The molecule shown in Figure 7.24 is an example of
   A A saturated alcohol
   B An unsaturated carboxylic acid
   C An unsaturated alcohol
   D A saturated carboxylic acid

Figure 7.24

4 Draw the full structural formula for pentan-3-ol.
5 Draw the full structural formula for propanoic acid.
6 Describe two uses for carboxylic acids.
7 Ethanol can be used as a fuel for cars. Write a chemical equation for the complete combustion of ethanol.
8 Describe two uses for esters.
9* Hydroxy acids are compounds that contain both a hydroxyl group and a carboxylic acid group within the same molecule. These compounds are able to form cyclic esters called lactones.

hydroxy acid   a lactone

Figure 7.25

   a) Draw the structural formula for the lactone formed when this hydroxy acid reacts.

Figure 7.26

   b) Draw the structural formula for the hydroxy acid from which this lactone was formed.

Figure 7.27

10* The structural formulae for two of the compounds in lavender oil are shown in Figure 7.28.

linalool   linalyl ethanoate

Figure 7.28

   a) Describe how bromine water could be used to prove that lavender oil contained unsaturated compounds.
   b) Linalool can be classed as an alcohol. Draw the part of the molecule that allows you to conclude that linalool is an alcohol.
   c) Linalyl ethanoate is made from an alcohol and ethanoic acid. To which group of compounds does linalyl ethanoate belong?
   d) Draw the full structural formula for ethanoic acid.

8 Energy from fuels

Alkanes and alcohols are excellent fuels (Figures 8.1 and 8.2). When fuels burn, they release lots of energy that we can use to heat our homes or power our cars. As chemists, we want to be able to compare fuels so that we know how much energy is released from burning a certain quantity of fuel. We will examine this idea in this chapter by using our knowledge of calculations from equations and we will also find out about a simple method for calculating the actual energy released from a fuel.

Figure 8.1 Cooking using a gas hob involves burning methane gas

Figure 8.2 Ethanol can be blended with petrol or diesel and used as a fuel for cars

Exothermic reactions

Chemical reactions that release energy are known as **exothermic reactions**. Lighting a match (Figure 8.3) is a good example of an exothermic reaction since it is obvious that energy is released in the form of heat.

Figure 8.3 Burning is an exothermic reaction

Figure 8.4 can be used to illustrate an exothermic reaction. It shows that the products have less energy than the reactants. Where has the energy gone? It has been released to the surroundings, mainly in the form of heat energy.

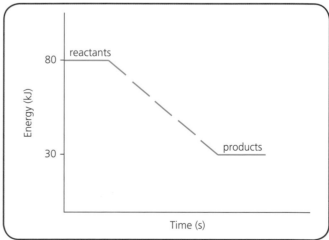

Figure 8.4 Exothermic reactions release energy to the surroundings

If we look at the values for energy on the *y*-axis, we can calculate how much energy has been released to the surroundings. The reactants had 80 kJ of energy at the start of the reaction and the products had 30 kJ at the end of the reaction. From these figures we can say that 50 kJ has been released to the surroundings. If this energy was released as heat, the temperature would rise. The word 'surroundings' can sometimes be a bit complicated so it is worth looking at an example of an exothermic reaction to understand exactly what is meant by the surroundings.

Think about adding some magnesium powder to hydrochloric acid (Figure 8.5).

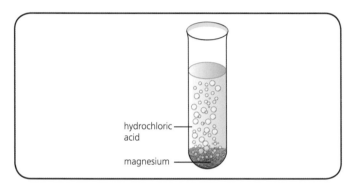

Figure 8.5 The temperature rises when magnesium is added to acid

If you have carried out this experiment in the laboratory, you will hopefully remember that the test tube becomes very warm. Here, heat energy is released to the surrounding acid, test tube and air – in other words, the surroundings are whatever is surrounding the reaction. If a thermometer was placed in the test tube, a rise in temperature would be measured.

Exothermic reactions are the most common reactions we encounter in our everyday lives as we tend to burn fuels for cooking and transport. Simple heat packs that are applied to the skin to soothe aching muscles and joints are another example of an exothermic reaction (Figure 8.6). The heat packs usually contain iron filings, salt and carbon. When exposed to air, the chemicals in the heat pack react and give out heat energy.

Figure 8.6 Heat packs release energy – an example of an exothermic reaction

Endothermic reactions

Some chemical reactions take in energy from the surroundings. These are known as **endothermic reactions**. For example, when barium hydroxide is mixed with ammonium chloride, the temperature falls as a reaction takes place (see Figure 8.7).

Figure 8.7 The temperature falls in an endothermic reaction

Figure 8.8 illustrates an endothermic reaction. You can see that the products have more energy than the reactants. Where has the energy come from? The energy has been taken away from the surroundings, so the surroundings become cooler. If we look at the scale on the *y*-axis, we can calculate that 50 kJ of energy has been removed from the surroundings. Endothermic reactions always take in energy from the surroundings.

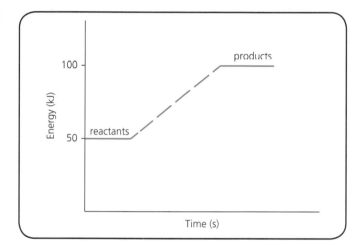

Figure 8.8 Endothermic reactions take in energy from the surroundings

Endothermic reactions are not as common as exothermic reactions, but we do find them in everyday life. For example, sherbet (Figure 8.9) contains sugar, citric acid and bicarbonate of soda. When the citric acid and bicarbonate of soda dissolve in your saliva and react with each other, there is a small drop in temperature.

Figure 8.9 Eating sherbet causes the temperature of your mouth to decrease – this is an endothermic reaction

Likewise, some cold packs (Figure 8.10) contain chemicals that react endothermically. These can be used to treat muscle sprains.

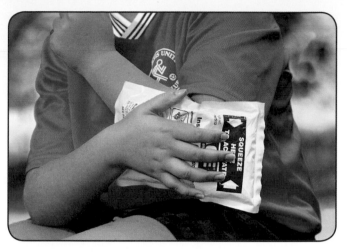

Figure 8.10 Some cold packs contain chemicals that react together to produce a drop in temperature

Writing combustion equations

Combustion (burning) is an obvious example of an exothermic reaction.

When chemicals burn they react with oxygen. For example, hydrogen gas burns readily to produce water:

$$H_2 + \tfrac{1}{2}O_2 \longrightarrow H_2O$$

When hydrocarbons burn in a good supply of oxygen, carbon dioxide and water are formed, as shown in the following equation:

$$\text{ethane} + \text{oxygen} \longrightarrow \text{carbon dioxide} + \text{water}$$
$$C_2H_6 + O_2 \longrightarrow CO_2 + H_2O$$

These equations can be balanced quite easily if you follow the 'CHO' rule: balance the carbon, then the hydrogen and, finally, the oxygen atoms. For burning ethane, the balanced chemical equation is:

$$C_2H_6 + 3\tfrac{1}{2}O_2 \longrightarrow 2CO_2 + 3H_2O$$

Alcohols also burn to produce carbon dioxide and water and the CHO rule can be applied to balance alcohol combustion equations too:

$$\text{methanol} + \text{oxygen} \longrightarrow \text{carbon dioxide} + \text{water}$$
$$CH_3OH + 1\tfrac{1}{2}O_2 \longrightarrow CO_2 + 2H_2O$$

Try the examples below to practise writing and balancing equations for the burning of hydrocarbons and alcohols. Review Unit 1 if you need extra help with balancing chemical equations.

1 Write balanced chemical equations for the complete combustion of:
   a) methane
   b) butane
   c) hexane.
2 Write balanced chemical equations for the complete combustion of:
   a) ethanol
   b) propanol
   c) hexanol.

Calculations from equations

In Unit 1 you learned about calculating quantities of reactants and products from balanced chemical equations. This skill can be applied to combustion equations where we are interested in calculating the quantities of reactants required or products formed.

3 Calculate the mass of water formed when 8 g of hydrogen is burned completely.
4 Calculate the mass of carbon dioxide formed when 60 g of ethane is burned completely.
5 Calculate the mass of water formed when 2.9 g of butane is burned completely.
6 Calculate the mass of carbon dioxide formed when 23 g of ethanol is burned completely.
7 Calculate the mass of ethanol required to produce 90 g of water when the ethanol is burned completely.
8 Calculate the mass of oxygen required to react completely with 17.6 g of propane.
9 Calculate the mass of butane required to react completely with 51.2 g of oxygen.
10 Calculate the mass of propanol required to produce 22 g of carbon dioxide.

Worked example

Calculate the mass of water produced when 32 g of methane is burned according to the following equation:

Step 1: Write a balanced equation, unless already given	$CH_4 \ + \ 2O_2 \ \longrightarrow \ CO_2 \ + \ 2H_2O$
Step 2: Identify the two chemicals referred to in the question and write the mole ratio	1 mole $\longrightarrow$ 2 moles
Step 3: Calculate the number of moles of the substance you have been given a mass for (methane)	$\text{number of moles} = \dfrac{\text{mass in grams}}{\text{mass of 1 mole}}$ $= \dfrac{32\,g}{16\,g}$ $= 2 \text{ moles}$
Step 4: Use the mole ratio to calculate the number of moles of the substance you are trying to find (water)	1 mole $\longrightarrow$ 2 moles 2 moles $\longrightarrow$ 4 moles
Step 5: Calculate the mass	mass in grams $= \begin{array}{c}\text{number} \\ \text{of moles}\end{array} \times \begin{array}{c}\text{mass of} \\ \text{1 mole}\end{array}$ $= \ \ 4 \ \ \times \ \ 18\,g$ $= 72\,g$

Calculating the energy released from fuels

In secondary school, a simple way of comparing the energy released from burning different foods (the fuel) is to burn the food under a test tube of water and to note the starting temperature and highest temperature reached. This is shown in Figure 8.11.

Figure 8.11 An experiment to compare the energy released when a food is burned

This experiment can provide a good comparison between different foods as long as some steps are taken to ensure that each test is fair. The volume of water heated must be the same in each experiment, the mass of food burned should be known and the distance between the burning food and test tube should be kept the same, which is quite tricky! When comparing liquid fuels, it is more convenient to use the experimental set-up shown in Figure 8.12.

Figure 8.12 An experiment to calculate the energy released when a fuel burns

In this experiment, the fuel is burned in a spirit burner and the heat released is used to heat the water in the beaker. The mass of fuel used to heat the water can be calculated by measuring the mass of the spirit burner at the start of the experiment and at the end of the experiment. Table 8.1 shows some sample data from an experiment to calculate the energy released when ethanol burns.

Mass of burner before burning	100.0 g
Mass of burner after burning	99.5 g
Initial temperature of water	20 °C
Highest temperature of water	50 °C
Volume of water	100 cm³

Table 8.1 Experimental data

Experiment 1: Burning ethanol

In order to convert the temperature rise into an energy value, we need to use an equation that relates the temperature rise to energy. The equation we will use is:

$$E_h = cm\Delta T$$

E_h = the energy released
c = the specific heat capacity of water
m = the mass of water heated in kilograms (we assume that 1 litre of water = 1 kg)
ΔT = the temperature change in °C

The **specific heat capacity** of water converts the temperature rise to an energy value. It is given the symbol c and has a value of $4.18\,kJ\,kg^{-1}\,°C^{-1}$. This value tells us that to raise the temperature of 1 kg of water by 1 °C requires 4.18 kJ.

It should make sense that to raise the temperature of an even larger mass of water would require much more energy (think about using a Bunsen to boil a small volume of water versus a large volume of water). For example, to raise the temperature of 10 kg of water by 1 °C would require ten times as much energy, in other words 41.8 kJ.

Similarly, to raise the temperature of 1 kg of water by a much greater amount would require more energy. For example, to raise the temperature of 1 kg of water by 2 °C would require $2 \times 4.18 = 8.36\,kJ$.

Let us apply this equation to the data in Table 8.1.

$$E_h = cm\Delta T$$

$c = 4.18$
$m = 0.1\,kg\ (100\,cm^3\ water = 100\,g = 0.1\,kg)$
$\Delta T = 50\,°C - 20\,°C = 30\,°C$
$E_h = 4.18 \times 0.1 \times 30 = 12.54\,kJ$

How much ethanol was burned? We can work this out by comparing the mass of the burner before and after burning: $100.0\,g - 99.5\,g = 0.5\,g$.

Overall we can conclude that 0.5 g of ethanol burns to release 12.54 kJ of energy.

If you were asked to calculate the energy released when, for example, 40 g of ethanol was burned, you could do this by using simple proportion:

Step 1: Write the relationship between mass and energy that you know	$0.5\,g \longrightarrow 12.54\,kJ$
Step 2: Calculate the energy for 1 g (12.54/0.5)	$1\,g \longrightarrow 25.08\,kJ$
Step 3: Calculate the energy for 40 g	$40\,g \longrightarrow$ $(40 \times 25.08\,kJ)$ $= 1003.2\,kJ$

Experiment 2: Burning methanol

An experiment was carried out to calculate the energy released when methanol burns. Use the data from Table 8.2 to show that methanol releases less energy than ethanol.

Mass of burner before burning	120.0 g
Mass of burner after burning	119.2 g
Initial temperature of water	29 °C
Highest temperature of water	54 °C
Volume of water	100 cm³

Table 8.2

$$E_h = cm\Delta T$$

$$E_h = 4.18 \times 0.1 \times 25 = 10.45\,kJ$$

Mass of methanol burned $= 120.0 - 119.2 = 0.8\,g$

In other words, 0.8 g of methanol released 10.45 kJ when burned.

$$0.8\,g \longrightarrow 10.45\,kJ$$

$$\frac{0.8\,g}{0.8} \longrightarrow \frac{10.45}{0.8}$$

$$1.0\,g \longrightarrow 13.06\,kJ$$

By comparing the energy released when 1 g of methanol burns to the energy released when 1 g of ethanol burns (from Experiment 1), you can see that much less energy is released by burning methanol.

Improving the experiment

When we compare the energy values calculated by our simple laboratory experiment to data tables, we find that our results suggest lower energies. This is because a lot of the heat produced by our burning fuels is released to the surroundings and does not heat the water. Also, if you have carried out this experiment you will have noticed that the flame from the burning fuels is often yellow and that the beaker containing the water quickly becomes coated in a layer of soot. This tells us that incomplete combustion is occurring. To improve this experiment, an instrument called a **bomb calorimeter** can be used, shown in Figure 8.13.

Figure 8.13 The bomb calorimeter

The calorimeter can be used to obtain much more accurate energy data as it is designed to overcome the two experimental problems that were discussed for the spirit burner:

- Heat loss to the surroundings: as the water completely surrounds the container where the combustion occurs, most of the heat generated will be transferred to the water.
- Incomplete combustion: the calorimeter is usually attached to an oxygen cylinder so that the sample is ignited in an oxygen-rich atmosphere. This ensures that the sample is completely burned.

Checklist for Revision

- I can state the definitions of the words 'exothermic' and 'endothermic' and I know that burning is an example of an exothermic reaction.
- I can write balanced chemical equations for the burning of hydrocarbons and alcohols.
- I can calculate quantities of reactants and products using chemical equations.
- I can calculate the energy released from a fuel using $E_h = cm\Delta T$.

End-of-chapter questions

1. For each of the following examples, state whether the reaction is exothermic or endothermic and calculate the energy released/taken in.
   a) A chemical reaction took place where the reactants had 80 kJ of energy. At the end of the experiment, the products were found to have 20 kJ of energy.
   b) A chemical reaction took place where the reactants had 120 kJ of energy. At the end of the reaction, the products were found to have 190 kJ of energy.
2. Calculate the mass of water formed when 20 g of hydrogen is completely burned.
3. Calculate the mass of carbon dioxide formed when 640 g of methane burns completely.
4. Calculate the mass of hexane required to produce 144 g of water on complete combustion.
5. Calculate the mass of carbon dioxide formed when 10 g of methanol is burned completely.
6. Calculate the mass of ethanol required to react completely with 100 g of oxygen.
7.* The energy released by burning methanol can be calculated using the apparatus shown in Figure 8.14. In an experiment, the results given in Figure 8.15 were obtained. Use these results to calculate the energy released when 32 g of methanol is burned.

Figure 8.14

mass of methanol burned = 0.45 g
temperature rise of water = 10 0 °C

Figure 8.15

8 Use the data in Table 8.3 to calculate the energy released when 100 g of propanol is completely burned using the apparatus shown in question 7.

Mass of burner before burning	120.0 g
Mass of burner after burning	119.2 g
Initial temperature of water	20 °C
Highest temperature of water	57 °C
Volume of water	100 cm³

Table 8.3

9* A student used the simple laboratory apparatus shown in Figure 8.16 to determine the energy released when methanol is burned.

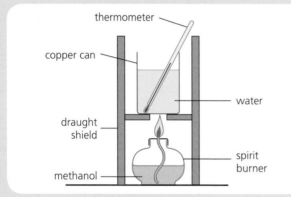

Figure 8.16

a) What measurements would have to be made to allow you to calculate the energy released by burning the methanol?

b) The student found that burning 0.370 g of methanol produces 3.86 kJ of energy. Calculate the energy released when 20 g of methanol is burned.

c) A more accurate value for the energy released by burning a fuel can be obtained using a bomb calorimeter (Figure 8.17). From your reading in this chapter, suggest why this experiment gives more accurate results than the simple spirit burner method.

Figure 8.17

10* Scientists have been experimenting to find ways of reducing carbon dioxide in the atmosphere. One of these ways involves placing concrete balls on the sea bed. They hope that green plants called algae will grow on the balls and this will help to reduce the carbon dioxide level. The balanced equation for the removal of carbon dioxide by algae is:

$$6CO_2 + 6H_2O \longrightarrow C_6H_{12}O_6 + 6O_2$$

If 200 kg of carbon dioxide were removed by the algae, calculate the mass of oxygen that would be produced.

Chemistry
in Society

9 Metals

Crystals and bonding

In Unit 1, you learned that ionic bonds are the electrostatic forces of attraction between positive and negative ions, and that these electrostatic attractions cause the oppositely charged ions to assemble themselves into a regular, three-dimensional structure called an **ionic lattice** Figure 9.1. The arrangement of ions in the lattice is a balance between the repulsion experienced by ions of the same charge and the attraction between ions of opposite charge.

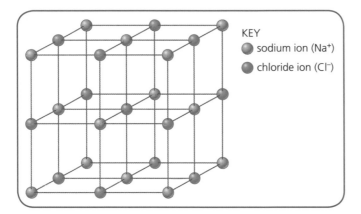

KEY
● sodium ion (Na⁺)
● chloride ion (Cl⁻)

Figure 9.1 An ionic lattice

Common salt contains sodium ions (Na⁺) and chloride ions (Cl⁻). Both of these ions have a fixed size (sodium ions are about half the size of chloride ions), so the ions assemble in a particular pattern; a pattern that repeats itself over and over again. This pattern is so regular that, as more and more ions join the lattice, **macroscopic** particles of a fixed shape (decided by the pattern) start to form and these become big enough for us to see. We call them **crystals**. Many substances that consist of a regular, three-dimensional array of particles can form crystals. For example, carbon atoms can arrange themselves into an **atomic lattice** and form crystals that we call diamonds. Odd though it may seem, metals also have a crystal structure.

Metallic bonding

Figure 9.2 We use metals in many ways

Metal atoms have a small number of outer electrons, so in order to achieve the electron arrangement of their nearest noble gas, metal atoms tend to lose electrons. When a metal atom loses its outer electron(s), a positive ion is formed. For example:

$$Mg \longrightarrow Mg^{2+} + 2e^-$$

(neutral   (positive (two electrons lost
atom)   ion)   by the Mg atom)

If we could see into the structure of a metallic element, we would see that all the metal atoms had lost their outer electrons and had formed a regular, three-dimensional array (lattice) of positively charged ions. These positive ions should all repel each other and blow the structure apart, so why does this not happen? The reason is that the electrons lost from the outer energy level of each atom remain in the lattice and move freely between the positive ions. This 'sea' of negative charge (electrons) floating through the **metallic lattice** does two things:

- It attracts the positive ions (holding them together).
- It balances the repulsion between the positive ions (stopping them being forced apart).

The free electrons in metals are usually said to be **delocalised** because they are not held to any one nucleus; they can move anywhere through the structure and the electrostatic attraction of the positive ions for the delocalised electrons is called **metallic bonding** (Figure 9.3).

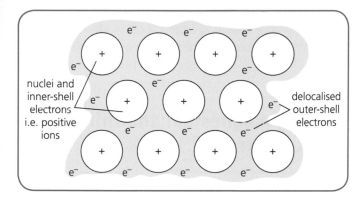

Figure 9.3 Metallic bonding

Properties of metals

Ionic lattices and metallic lattices both contain charged particles, but in an ionic lattice, the oppositely charged ions in the compound have different sizes and different charges, so they cannot change places without destroying the pattern of the lattice. A metal, however, is an element, so in a pure metal, the positive ions are all identical; they are all the same size and they can change places in the lattice and slide over each other without breaking the structure apart. This gives metals some very useful properties. For instance, if you pick up a small copper(II) sulfate crystal (taking care to wear gloves) and try to bend it, it will break, but if you do the same

to a small piece of metal foil, it will not break; it will bend. We say metals are **malleable**, which means they can be shaped by beating them with a hammer. This property has allowed craftsmen to make tools (and swords!) from metal for thousands of years.

Metallic bonding also makes metal elements very good conductors of electricity. Electricity is a flow of charged particles and, if a voltage is applied across a piece of metal, the free (delocalised) electrons can move easily through the lattice. As a result, an electric current can flow through the metal without any chemical change taking place. Metals and graphite both conduct electricity in the solid state because their structures contain delocalised electrons. Due to their type of bonding, metals can conduct electricity when they are solid and when they are **molten** (in the liquid state) and they always remain chemically unchanged by the electron flow.

It is important to remember that although the delocalised electrons in a metal are no longer held by one particular nucleus, they have not been transferred to another type of atom; the metal has not reacted with anything and the outer electrons from each nucleus are all still present in the lattice. This means that although the lattice is an array of positive ions, the metal element consists of atoms. The symbol for the element magnesium is Mg. The symbol Mg^{2+} is only used when magnesium atoms have reacted with another type of atom and the element, magnesium, has become part of a compound.

Activities

Farriers make iron horseshoes to attach to horses' hooves. They use a hammer to beat the hot metal into shape. Use your knowledge of chemistry to write a short paragraph describing what happens to the metal lattice when a horseshoe is being forged (being made) and explain why the piece of iron does not break when the hammer hits it.

1 a) What is metallic bonding?
  b) What is meant by the word 'malleable'?
2 'Metallic bonding has some of the characteristics of covalent bonding and some of the characteristics of ionic bonding.' Explain whether you agree or disagree with this statement, giving reasons for your answer.

3 a) Give the name and symbol for a metal which is not a solid at room temperature (25 °C).
  b) In chemistry, what does the word 'lattice' mean?
  c) Describe what happens when a piece of copper is hit with a hammer. Explain why this happens.

Reactions of metals

Metal atoms all have relatively small numbers of outer electrons, so they have one chemical property in common: when they take part in a chemical reaction, they tend to lose electrons. We often say that metal atoms tend to give electrons away or 'donate' electrons. The atoms of some metals lose (donate) electrons more easily than others. We say that metals whose atoms lose electrons easily are **reactive metals** and metals whose atoms do not lose electrons easily are **unreactive metals**. Due to this tendency to give electrons away, all metals have similar reactions with air (oxygen), water and with dilute acids. It is possible to put together a reactivity series of metals based on their reactions with oxygen, water and dilute acids. This series is shown in Table 9.1.

Name	Symbol
potassium	K
sodium	Na
lithium	Li
calcium	Ca
magnesium	Mg
aluminium	Al
zinc	Zn
iron	Fe
tin	Sn
lead	Pb
copper	Cu
mercury	Hg
silver	Ag
gold	Au

Table 9.1 Some metals are more reactive than others. This table gives a reactivity series for some common metals, with the most reactive metals at the top and the least reactive at the bottom

Activities

If there is time, your teacher/lecturer may give you some iron filings and a small piece of copper foil. Wet the end of a wooden spill and dip it into the iron filings, then hold the spill at the edge of a blue Bunsen flame for a few seconds. Use a pair of tongs to hold the piece of copper foil in the edge of a blue Bunsen flame for 60 seconds then place it on a heatproof mat. Record your observations in a table. Explain why this experiment will not allow you to draw a valid conclusion about the relative reactivity of these two metals.

Burning magnesium and thinking about electrons

Activities

Demonstration: if time allows, your teacher/ lecturer may use a pair of tongs to hold a piece of magnesium in a Bunsen flame until it ignites. Take care not to look directly at the burning magnesium: look to the left or right of it. All that energy is being released simply because electrons are moving from one type of atom to another.

The balanced equation for the **combustion** of magnesium is:

$$2Mg(s) + O_2(g) \longrightarrow 2MgO(s)$$

When metals take part in a chemical reaction, they lose electrons. In this reaction, the electrons lost (as the magnesium atoms become magnesium ions, Mg^{2+}) must be gained by the atoms in the oxygen molecule, which means the oxygen atoms become oxide ions, O^{2-}.

Equation 1: $2Mg(s) + O_2(g) \longrightarrow 2Mg^{2+}O^{2-}(s)$

We can split this equation into two halves; one half shows what happens when a magnesium atom (electron arrangement 2,8,2) loses its two outer electrons:

Equation 2: $Mg \longrightarrow Mg^{2+} + 2e^-$

The other half shows what happens when an atom from the oxygen molecule (the electron arrangement of an oxygen atom is 2, 6) gains the electrons lost from the magnesium atom:

Equation 3: $\frac{1}{2}O_2 + 2e^- \longrightarrow O^{2-}$

Equations 2 and 3 each represent half of the reaction shown in equation 1 and since they contain both ions and electrons, they are called **ion–electron half equations** (often simply called ion–electron equations). When magnesium burns, it combines with oxygen and we say that it has been **oxidised**. Equation 2 shows that when magnesium burns, its atoms lose electrons, so we can also say that when a metal atom loses electrons, it is being oxidised. In chemistry, the opposite of oxidation is reduction, so we say the oxygen atoms in equation 3 are being **reduced**, in other words they are gaining electrons. An easy way to remember how the electrons move in these two processes is to use the mnemonic OIL RIG:

Oxidation **I**s **L**oss (of electrons); **R**eduction **I**s **G**ain (of electrons)

Remember: in an ion–electron equation showing oxidation (often indicated by writing [O] following it), the electrons are moving away (being lost) from the reactant, so the electrons appear on the *right* of the arrow, but in an ion–electron equation showing reduction (often indicated by writing [R] following it), the electrons are moving towards (being gained by) the reactant, so the electrons appear on the *left* of the arrow.

Reaction of metals with oxygen

All the metals above silver in the reactivity series (see Table 9.1) combine with oxygen when heated. The reaction produces a metal oxide. The higher the metal is in the series, the more violent the reaction between the metal and oxygen. In general, this reaction takes the form:

$$metal + oxygen \longrightarrow metal\ oxide$$

For example,

$$copper + oxygen \longrightarrow copper(II)\ oxide$$
$$2Cu + O_2 \longrightarrow 2CuO$$

Metal oxides are ionic compounds, so have high melting points and are solids at room temperature. Metallic bonding is usually strong, so most metal elements have got high melting points and all but one of the metal elements exist in the solid state at room temperature (25°C).

Showing ions and state symbols, the equation becomes:

$$2Cu(s) + O_2(g) \longrightarrow 2Cu^{2+}O^{2-}(s)$$

The apparatus shown in Figure 9.4 can be used to study the reactivity of metals from magnesium down to copper in the reactivity series.

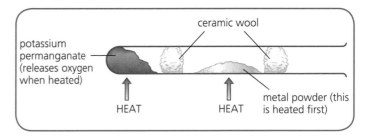

Figure 9.4 Metals glow brightly as they react with oxygen

The three metals at the top of the reactivity series – potassium, sodium and lithium – are so reactive that they are stored under oil to prevent them from reacting with the oxygen and water in the air.

Reactions of metals with water

The metals above aluminium in the reactivity series react with cold water to produce hydrogen gas and the corresponding metal hydroxide. The general equation for this is:

$$metal + water \longrightarrow metal\ hydroxide + hydrogen$$

For example:

$$sodium + water \longrightarrow sodium\ hydroxide + hydrogen$$
$$2Na + 2H_2O \longrightarrow 2NaOH + H_2$$

If we rewrite this equation to show ions and state symbols, we get:

$$2Na(s) + 2H_2O(l) \longrightarrow 2Na^+(aq) + 2OH^-(aq) + H_2(g)$$

Reactions of metals with dilute acids

All metals above copper in the reactivity series react with dilute acids such as hydrochloric acid and sulfuric acid to produce a salt and hydrogen:

$$metal + acid \longrightarrow salt + hydrogen$$

For example:

zinc + hydrochloric acid $\longrightarrow$ zinc(II) chloride + hydrogen

$$Zn + 2HCl \longrightarrow ZnCl_2 + H_2$$

If we rewrite this equation to show ions and state symbols, we get:

$$Zn(s) + 2H^+(aq) + 2Cl^-(aq) \longrightarrow$$
$$Zn^{2+}(aq) + 2Cl^-(aq) + H_2(g)$$

When a metal reacts with an acid it produces bubbles of hydrogen gas. In general, the faster the bubbles are produced, the more reactive the metal. One exception is aluminium. It reacts slowly with acids for about the first 20 minutes, after which it reacts quickly. The reason for this is that the aluminium is protected by a coating of aluminium oxide, which must be removed by the acid before the metal can react. Table 9.2 summarises the reactions of metals.

Supply and demand

Humans have been digging up metal **ores** and extracting the metals for thousands of years. A metal ore is a rock that contains a metal compound (miners call metal compounds 'minerals') and a good knowledge of chemistry is needed to get (extract) the pure metal from each metal compound. A lot of mines are now worked out, which means that the amount of mineral left in the mine is not worth the cost of digging it up, and the big mining companies around the world are constantly looking for new supplies. We know about the demand for metals such as iron, aluminium and copper because we see them in use every day, but there is a group of

| Metal | Reaction with | | |
	Oxygen	Water	Dilute acid
potassium	metal + oxygen ↓ metal oxide	metal + cold water ↓ metal hydroxide + hydrogen	metal + acid ↓ salt + hydrogen
sodium			
lithium			
calcium			
magnesium			
aluminium		no reaction	
zinc			
iron			
tin			
lead			
copper			no reaction
mercury			
silver	no reaction		
gold			

Table 9.2

elements called the **rare earth metals** which are in high demand because they are essential for the manufacture of computers, mobile phones and GPS satellites. China currently supplies around 90% of the world production of rare earth metals but, if we do not conserve supplies, we may run out of the raw materials needed for our complex modern technologies.

Questions

4 a) Write a balanced equation for the reaction between potassium and oxygen.
   b) Write an equation showing the ions present in the reaction. Use the equations given in the text as a guide.
5 For the reaction between calcium and water, give:
   a) a balanced equation
   b) an equation showing ions and state symbols.
6 For the reaction between magnesium and dilute hydrochloric acid, give:
   a) a balanced equation
   b) an equation showing ions and state symbols.

7 A, B and C are three metals. Metal A reacts with dilute hydrochloric acid but not with water. Metal B does not react with water or dilute acid. Metal C reacts with water and dilute acid.
   a) Place the metals in order of reactivity, with the most reactive first.
   b) Using the reactivity series in Table 9.2, give all the possible metals that A, B and C could be.

Figure 9.5 Modern boats use hi-tech radar equipment

Questions

8 The semiconductor gallium nitride is essential for high-power radar systems (see Figure 9.5), so trade in gallium nitride is restricted.
   a) Which group of the Periodic Table is gallium in?
   b) What is the formula for gallium nitride?
   c) Name another material that is more commonly used as a semiconductor.
9 Name two rare earth metals and give the symbol for each of them. (You may need to carry out some internet research to answer this question.)

Extraction of metals

Gold (Figure 9.6) is a very unreactive metal (it tends not to form compounds with other elements) so it can still be found as an element in the Earth's crust. Gold is

Figure 9.6 Gold jewellery that dates from the Bronze Age

almost always found as a natural alloy, mixed with other elements, usually silver. There are some deposits of other uncombined metals around the world, but most metals are obtained from their naturally occurring compounds.

Industrial chemists have developed complex ways to extract some metals, but the basic chemistry used to extract the more common metals has been unchanged for hundreds, if not thousands, of years. The first step is to crush the ore into small pieces, then heat it in air to form the metal oxide. The oxygen is then removed by a process called reduction and the metal is obtained.

The reactivity series shown in Table 9.2 can be used to decide how easy it is to extract a metal from its oxide. Metals towards the bottom of the series (unreactive metals) do not form compounds easily, so it does not require much energy to break up (**decompose**) their oxides and get the pure metal. Metals towards the top of the series (reactive metals) form compounds easily, so it requires a great deal of energy to extract those metal elements from their oxides. For this purpose, we can split the reactivity series at two points producing three groups of metals.

1 Metals below copper

The oxides of these metals decompose when they are heated to give the metal and oxygen.

$$\text{metal oxide} \longrightarrow \text{metal} + \text{oxygen}$$

For example:

$$\text{silver(I) oxide} \longrightarrow \text{silver} + \text{oxygen}$$

$$2Ag_2O \longrightarrow 4Ag + O_2$$

Showing ions and state symbols, the equation becomes:

$$2(Ag^+)_2O^{2-}(s) \longrightarrow 4Ag(s) + O_2(g)$$

2 Metals above mercury and below aluminium

The oxides of these metals do not decompose when they are heated, but if the oxide is mixed with carbon and then heated, the carbon and oxygen combine to give carbon dioxide and leave the metal as an element. This removal of oxygen is called **reduction**. Carbon is doing the reducing (removing oxygen) so carbon is described as a **reducing agent**.

metal oxide + carbon $\longrightarrow$ metal + carbon dioxide

For example:

zinc(II) oxide + carbon $\longrightarrow$ zinc + carbon dioxide

$$2ZnO + C \longrightarrow 2Zn + CO_2$$

Showing ions and state symbols, the equation becomes:

$$2Zn^{2+}O^{2-}(s) + C(s) \longrightarrow 2Zn(s) + CO_2(g)$$

3 Metals above zinc

Compounds of these metals are so stable (so hard to decompose) that electricity has to be used to obtain the metal. Metals above zinc in the reactivity series can only be obtained by **electrolysis** of their molten compounds. Aluminium is obtained from its molten oxide, but the rest are obtained from their chlorides. As the compound is heated, the ionic lattice breaks down and the ions become free to move. During electrolysis, positive metal ions are attracted to the negative electrode where they gain electrons and form metal atoms. For example:

$$Al^{3+}(l) + 3e^- \longrightarrow Al(l) \ [R]$$

We say metal ions are reduced at the negative electrode to form metal atoms.

Percentage composition

Most metals are found in the Earth's crust as different ores. For example, two different iron ores are haematite and magnetite, both of which contain oxides of iron. When industrial chemists set up an extraction process, they need to know how much metal they can expect to obtain from each tonne of ore, so they need to know how to calculate the percentage mass composition of a compound. Before we start thinking about this, we will revise some earlier work covered in Unit 1.

Relative formula mass

The formula mass of a compound is found by adding together the **relative atomic masses** of each atom in the formula. Values for the relative atomic masses of some elements can be found in the data booklet. For example, to calculate the relative formula mass of sodium hydroxide:

formula is NaOH

formula mass is 23 + 16 + 1 = 40

If more than one atom of an element is present in the formula, the relative atomic mass of the element is

Activities

1 Use the internet to find out more about the extraction of aluminium from its ore. Think about the following questions:
 - What is the name of the main ore of aluminium?
 - How is the metal extracted?
 - Why is the ore mixed with cryolite?
 - What material is used to make the electrodes?
 - Why does the positive electrode need to be replaced more often than the negative electrode?

2 Since the start of the twentieth century, aluminium has been relatively cheap and plentiful, so why did Napoleon Bonaparte have the state dinner plates made from aluminium as a sign of his great wealth?

Questions

10 For the effect of heat on mercury(II) oxide
 a) write a balanced equation
 b) write an equation showing ions and state symbols.
11 Copper can be obtained by heating copper(II) oxide with carbon. For this reaction:

 a) write a balanced equation
 b) write an equation showing ions and state symbols
 c) name the reducing agent.

multiplied by the appropriate number. For example, for lithium carbonate:

formula is Li_2CO_3

formula mass is $(2 \times 7) + 12 + (3 \times 16) = 74$

If the formula of the compound contains ions that contain more than one type of atom (usually called group ions), the mass of every atom inside the bracket is multiplied by the number outside the bracket. For example, for aluminium sulfate:

formula is $Al_2(SO_4)_3$

formula mass is $(2 \times 27) + (3 \times 32) + (12 \times 16) = 342$

Hint: to avoid unforced arithmetic errors, it is often safer to calculate the mass of the atoms inside the bracket then multiply that mass by the number outside the bracket:

formula is $Al_2(SO_4)_3$

formula mass is $(2 \times 27) + [32 + (4 \times 16)] \times 3$

$= (2 \times 27) + [96] \times 3$

$= 342$

Once we have calculated the relative formula mass of a compound, we can find the percentage by mass of any element in the compound by using the equation given below:

$$\text{percentage mass of a given element in a compound} = \frac{\text{mass of element in the formula}}{\text{formula mass}} \times 100$$

Worked examples

1 The formula mass of haematite (Fe_2O_3)
   $= (2 \times 56) + (3 \times 16)$
   $= 160$

   The percentage of iron in haematite $= \dfrac{\text{mass of iron in formula}}{\text{relative formula mass}} \times 100$

   $= \dfrac{(2 \times 56)}{160} \times 100$

   $= 70\%$

2 The percentage of oxygen in aluminium sulfate, $Al_2(SO_4)_3$

   $= \dfrac{\text{mass of oxygen in formula}}{\text{relative formula mass}} \times 100$

   $= \dfrac{(12 \times 16)}{342} \times 100$

   $= 56.14\%$

Questions

12 Calculate the percentage by mass of iron in the ore magnetite, formula Fe_3O_4.
13 Calculate the percentage by mass of aluminium in the compound alumina. This can be assumed to be pure aluminium oxide, Al_2O_3.

14 Calculate the percentage by mass of:
   a) sodium in sodium chloride (NaCl)
   b) lead in lead(II) bromide $(PbBr_2)$
   c) zinc in zinc(II) carbonate $(ZnCO_3)$
   d) calcium in calcium phosphate $(Ca_3(PO_4)_2)$.

Reduction, oxidation and redox reactions

Whenever a reaction involves both oxidation and reduction, it is called a **redox reaction**. In a redox reaction, electrons lost by one substance during oxidation are gained by another substance during reduction. Three examples of redox reactions involving metals are their reactions with water, oxygen and dilute acids.

Metal + water

Sodium reacts readily with water according to the equation:

sodium + water $\longrightarrow$ sodium hydroxide + hydrogen

overall reaction:

$2Na(s) + 2H_2O(l) \longrightarrow 2Na^+OH^-(aq) + H_2(g)$

We can split this redox equation up into ion–electron equations to show how electrons are transferred in the redox reaction. Sodium atoms are oxidised (lose electrons):

oxidation part: $Na(s) \longrightarrow Na^+(aq) + e^-$

Hydrogen ions present in water are reduced (gain the electrons lost by the sodium atoms):

reduction part: $2H^+(aq) + 2e^- \longrightarrow H_2(g)$

From the balanced equation we can see that two sodium atoms reacted. This was to produce the two electrons required to reduce hydrogen ions to hydrogen gas, so we can write equations to show the number of electrons transferred:

$$2Na(s) \longrightarrow 2Na^+(aq) + 2e^- \text{ [O]}$$

$$2H^+(aq) + 2e^- \longrightarrow H_2(g) \text{ [R]}$$

So the **redox equation** for this reaction would be:

$$2Na(s) + 2H^+(aq) \longrightarrow 2Na^+(aq) + H_2(g)$$

Note that this redox equation does not include the hydroxide ions present in water (reactant) or the hydroxide ions of the sodium hydroxide (product). Redox equations only show the substances that lost and gained electrons.

Metal + oxygen

Calcium reacts readily with oxygen molecules in the atmosphere according to the equation:

calcium + oxygen $\longrightarrow$ calcium oxide

overall reaction:

$$2Ca(s) + O_2(g) \longrightarrow 2Ca^{2+}O^{2-}(s)$$

We can split this redox equation up into ion–electron equations to show how electrons are transferred in the redox reaction. Calcium atoms are oxidised (lose electrons):

oxidation part: $Ca(s) \longrightarrow Ca^{2+}(s) + 2e^-$

Oxygen atoms are reduced (gain the electrons lost from the calcium atoms):

reduction part: $\frac{1}{2}O_2(g) + 2e^- \longrightarrow O^{2-}(s)$

From the balanced equation we can see that two calcium atoms are needed to react with each oxygen molecule, so we can write equations to show the number of electrons transferred in each reaction:

$$2Ca(s) \longrightarrow 2Ca^{2+}(s) + 4e^- \text{ [O]}$$

$$O_2(g) + 4e^- \longrightarrow 2O^{2-}(s) \text{ [R]}$$

Note: If the ion–electron equations for a reaction contain equal numbers of electrons, they are simply added together to give the overall redox equation with the electrons omitted.

Redox equations must *never* contain electrons. The following is an incorrect redox equation:

$$2Ca(s) + O_2(g) + 4e^- \longrightarrow 2Ca^{2+}O^{2-}(s) + 4e^-$$

The electrons lost and gained must be left out to give the following correct redox equation:

$$2Ca(s) + O_2(g) \longrightarrow 2Ca^{2+}O^{2-}(s)$$

Metal + dilute acid

The reaction between a metal and a dilute acid is a redox reaction because the metal atoms lose electrons to form metal ions and the hydrogen ions, present in the acid, gain electrons to form hydrogen gas. For example, magnesium reacts with dilute hydrochloric acid:

magnesium + hydrochloric acid $\longrightarrow$
magnesium chloride + hydrogen

$$Mg + 2HCl \longrightarrow MgCl_2 + H_2$$

Equation showing ions and state symbols:

overall reaction:

$$Mg(s) + 2H^+(aq) + 2Cl^-(aq) \longrightarrow$$
$$Mg^{2+}(aq) + 2Cl^-(aq) + H_2(g)$$

oxidation part: $Mg(s) \longrightarrow Mg^{2+}(aq) + 2e^-$

reduction part: $2H^+(aq) + 2e^- \longrightarrow H_2(g)$

If we add these two ion–electron equations and cancel the electrons, we get the following redox equation:

$$Mg(s) + 2H^+(aq) \longrightarrow Mg^{2+}(aq) + H_2(g)$$

There are no chloride ions in this equation because the chloride ions, $Cl^-(aq)$, are neither oxidised nor reduced during the reaction. The chloride ions are present in the mixture, but they play no part in the reaction. They are said to be **spectator ions** (see Unit 1, Chapter 5). Redox equations only show the substances that are involved in the oxidation part and the reduction part of the reaction; spectator ions are never included in redox equations.

Activities

Your teacher/lecturer may demonstrate the experiment shown in Figure 9.7. Use the electrochemical series from the data booklet to help you write ion–electron equations for the formation of lead metal at the negative electrode and the formation of bromine at the positive electrode. Add these two equations together and write the overall equation for the decomposition of lead(II) bromide. Explain how these equations tell you that electrolysis is a redox process. Are there any spectator ions in this reaction?

Figure 9.7 Passing a direct current through molten lead(II) bromide

Questions

15 State whether the following are oxidation or reduction reactions:
   a) $Na \longrightarrow Na^+ + e^-$
   b) $I_2 + 2e^- \longrightarrow 2I^-$
   c) $S + 2e^- \longrightarrow S^{2-}$.

16 Identify the redox reaction shown in the grid.

A	B
$2F^- \longrightarrow F_2 + 2e^-$	$Fe + Pb^{2+} \longrightarrow Fe^{2+} + Pb$
C	D
$e^- + K^+ \longrightarrow K$	$2O^{2-} \longrightarrow O_2 + 4e^-$

17 In the first stage of rusting, iron atoms lose electrons to form iron(II) ions:

Equation 1: $Fe \longrightarrow Fe^{2+} + 2e^-$

The electrons lost by the iron are gained by water and oxygen molecules:

Equation 2: $2H_2O(l) + O_2(g) + 4e^- \longrightarrow 4OH^-(aq)$

   a) Copy the ion–electron equations 1 and 2 and indicate which one represents reduction [R] and which represents oxidation [O].
   b) Adjust the number of electrons lost and gained in equations 1 and 2, then add the two ion–electron equations to obtain a balanced redox equation.

Electrochemical cells

When two different metals are connected by an **electrolyte**, an electric current flows from one metal to the other through connecting wires (Figure 9.9). Comparing pairs of metals allows us to construct a

Figure 9.8 Handsets such as this are battery operated

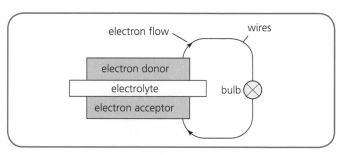

Figure 9.9 How an electrochemical cell works

list called the **electrochemical series**, which can be used to predict the size of the voltage between any two connected metals and the direction of current in chemical **cells**. The bigger the gap, the bigger the voltage and electrons flow from the metal higher in the series to the metal lower in the series. The electrochemical series can be found in the data booklet.

Figure 9.10 shows that electrons are being transferred through the wires (often called the 'external circuit') from zinc (the metal higher in the electrochemical series) to copper (the metal lower in the electrochemical series). This means that in the zinc half-cell, zinc atoms are being oxidised to form ions, and in the copper half-cell, copper ions are being reduced to form atoms. Ion–electron equations for these reactions are:

oxidation: $Zn(s) \longrightarrow Zn^{2+}(aq) + 2e^-$

reduction: $Cu^{2+}(aq) + 2e^- \longrightarrow Cu(s)$

These two equations contain equal numbers of electrons so they can be added (remembering to omit the electrons) to give the redox equation:

$$Zn(s) + Cu^{2+}(aq) \longrightarrow Zn^{2+}(aq) + Cu(s)$$

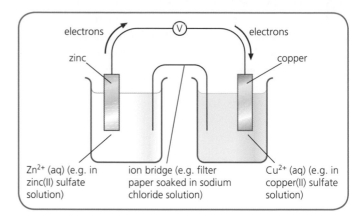

Figure 9.10 A chemical cell made by joining copper and zinc half-cells

Questions

18 In a Zn/Cu cell, electrons flow from the zinc to the copper through the wires and meter. Use the data booklet to give the direction of electron flow in each of the following cells:
   a) Mg/Fe
   b) Ag/Sn
   c) Ni/Zn
   d) Al/Pb.

19 For each of the following, state which cell would give the larger voltage:
   a) Ni/Pb and Fe/Cu
   b) Fe/Ag and Sn/Ag
   c) Al/Hg and Mg/Au
   d) Zn/Pb and Al/Hg.

Activities

This activity involves a study of 'fruity cells' that use a fruit and two different metals to produce an electric current. There are two parts to this investigation, but you will only be asked to work on one.

1 Find out which fruit makes the best electrochemical cell. You need to plan, carry out and then write a brief report on an experiment to find which fruit gives the highest voltage when it is used in a cell.

2 Find out which pair of metals (from a selection provided by your teacher) produces the highest voltage when used to make a fruity cell. You will need to plan, carry out and write up your experiment.

Decide whether you are going to do Experiment 1 or 2 then read the following advice before starting the task.

First, write down an aim for your experiment; this is what you are trying to do.

Think about the variables involved in the reaction, then write a plan to show what you intend to do and decide how you are going to record the observations.

It is usually a good idea to draw a diagram (in pencil) of the apparatus you are going to use and to make a list of the chemicals and apparatus you will need. Think about how you are going to ensure your experiments are safe. Ask your teacher/lecturer to check your plan.

When you have finished the practical work, record your results and write a 'Conclusion' (what you have found out). Always try to make the conclusion relate to the aim of your experiment.

Think about ways in which you could make the experiment more accurate. If you were going to do the experiment again, would you have done everything in the same order? Did you make any mistakes? Thinking about the answers to these questions, write one or two sentences to evaluate your experimental procedures.

Electrochemical cells as redox reactions

All electrochemical cells involve the movement of electrons from one half-cell to the other; this means that all electrochemical cells involve redox reactions. Cells are portable devices that convert chemical energy into electrical energy and in the past 50 years, a lot of work has been put into improving their design and efficiency. In many modern cells there are no metal elements involved in the redox reaction. A cell can be constructed from any chemical reaction that involves oxidation and reduction, for instance, the reaction between iron(III) chloride solution and potassium iodide solution. The two solutions can be connected using an ion bridge as before, but the electrodes can be graphite – a conducting non-metal. Electrons flow from the iodide ions in one solution through the meter to the iron(III) ions in the other. As this happens, the iodide ions are oxidised and turn into iodine molecules:

$$2I^- \longrightarrow I_2 + 2e^- \text{ [O]}$$

The iron(III) ions gain electrons (are reduced) and turn into iron(II) ions:

$$\text{(yellow) } Fe^{3+} + e^- \longrightarrow Fe^{2+} \text{ (green) [R]}$$

Balancing the electrons lost and gained (by multiplying the reduction reaction by two) gives:

$$2Fe^{3+} + 2e^- \longrightarrow 2Fe^{2+}$$

This equation can now be added to the oxidation equation (remembering to omit the electrons) to give the balanced redox equation.

$$2Fe^{3+} + 2I^- \longrightarrow 2Fe^{2+} + I_2$$

The spectator ions in this reaction are chloride ions and potassium ions.

Electrodes

If we make a cell by joining two different metals together, then the metals are used to conduct electricity to and from the external circuit, in other words the metals are used as electrodes. Figure 9.11 shows the cell referred to above in which neither of the electrodes is a metal. Electrons flow from one solution to the other, and so the electrodes can be any material that conducts

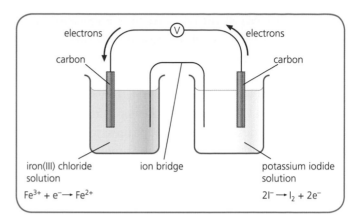

Figure 9.11 An iodide/iron(III) cell

electricity and does not react with the solution in which it has been placed. Graphite is often used for this purpose.

The next step?

Normally heart pacemakers need new batteries every 7 years and that requires drastic, chest-opening surgery. Aerospace engineers in the USA have developed a prototype made from magnets and a **ceramic piezoelectric** material which converts vibrations in the chest cavity (heartbeats) into electricity. At the time of writing (2012), live testing still needs to be done, but the engineers have used similar devices successfully to make small-scale energy 'harvesters' for spaceships and they say their system is small enough to fit inside the present pacemaker battery casing.

Rechargeable batteries

What we call an AA 'battery' should really be called a 'cell'. Batteries are only formed when two or more cells are connected together. Cells are little packages of useful chemicals that can produce electricity. The problem with cells is that they are sealed, so they contain a fixed quantity of reactants; when the reactants are used up in any of its cells, a battery runs down. A normal zinc–carbon cell (invented over 100 years ago by Leclanché) provides 1.5 volts when it is new, but as the cell is used, the voltage decreases rapidly until the power is too low to be useful. Lithium cells provide a more consistent voltage for longer, but a lot of recent research and development has gone into rechargeable cells.

In a rechargeable cell, the redox reaction is reversible. If a rechargeable cell goes flat (because the reactants have been used up), all we need do is pass a mains electric current through it and the redox reaction is forced into reverse; the original reactants are thus recreated and the cell can be used all over again. Different rechargeable cells have been designed for different uses: some have to provide a steady current for a long time; others need to keep their charge even when they are not being used; some need to be very lightweight; others have to work in very hot or very cold conditions. Two common batteries which contain rechargeable cells are lead–acid batteries and silver–zinc batteries. Lead–acid batteries are most commonly found in cars (Figure 9.12). They provide a high voltage for a low cost, but are very heavy and cannot hold their charge for long if they are not used regularly. Silver–zinc cells are found in hearing aids. They give a high current, high voltage and have the same watt-hour capacity as six lead–acid batteries, but they are expensive.

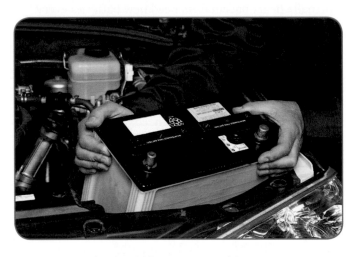

Figure 9.12 A lead–acid battery as used in a car

If there is time, your teacher/lecturer may show you how a lead–acid cell works.

low-voltage supply (6 V)

bulb lights up

charge cell for 5 minutes

lead strips   dilute sulfuric acid

Figure 9.13 A simple lead–acid cell

Activities

Questions

20 **a)** What is the purpose of an electrolyte in an electrochemical cell?
 b) Why do batteries only last for a certain length of time?
21 Each lead–acid cell can deliver about 2 volts when charged. The average car battery delivers about 12 volts. How many lead–acid cells are in the average car battery?

Fuel cells

Rechargeable batteries have a finite life; they can only be charged up and run down a certain number of times before they have to be replaced. In addition, the manufacture of batteries requires energy, causes pollution and uses up metals that are precious natural resources.

In contrast, hydrogen fuel cells use hydrogen and oxygen to make electricity and water (Figure 9.14).

$\frac{1}{2}O_2$

H_2

$2H^+$

H_2O

$2e^-$

Figure 9.14 A hydrogen fuel cell. This is not burning; very little heat is produced in the reaction and almost all the energy released is collected as electricity

The idea of this apparently simple and elegant device was suggested in 1839 by a High Court judge (and amateur scientist) called William Grove, but it was not until 100 years later that engineers made the first practical fuel cells. Worldwide concern over CO_2 in the atmosphere and the determination of the US State and Federal Governments to produce zero-emission vehicles is driving fuel cell research. Modern cells are achieving 70% conversion of chemical to electrical energy.

Average data show that for 2 days out of every 3, wind turbines do not produce electricity. If we are to keep the lights on in the future, every kilowatt of electricity produced by a wind turbine must have a matching kilowatt from a conventional or nuclear power station waiting on the national grid in case the wind drops. Putting this reality together with the fact that by 2050 cheap oil will no longer be available and Europe's own reserves will have been used up, we need a way to convert fuel gases into electricity

without also producing polluting gases. Scientists and engineers are doing some amazing work and there are now cells that use commonly available fuels such as carbon monoxide, biogas and even MDF waste (from the manufacture of furniture) at a level of efficiency that promises to reduce CO_2 emissions, in addition to the hydrogen fuel cells that produce almost zero emissions.

Checklist for Revision

- I understand and can explain metallic bonding theory.
- I can explain how an electric current flows in elements and compounds.
- I understand some typical reactions of metals and can explain how they occur in terms of electron flow, redox reactions, oxidation and reduction.
- I know how metals are extracted from ores by reduction reactions.

- I can write balanced ionic equations for the reactions of metals and the extraction of metals from ores.
- I can explain how electrochemical cells work, including those that utilise non-metal electrodes.
- I understand how rechargeable batteries work.
- I understand modern cell technology, including fuel cells.

End-of-chapter questions

1 Lead can be obtained by heating lead(II) oxide with charcoal (a form of carbon).
   a) Write a balanced symbol equation for the reaction between lead(II) oxide and carbon.
   b) Suggest why magnesium cannot be obtained from its oxide by heating with carbon.
   c) Name a metal that can be obtained from its oxide by heating alone.

2 a) Suggest a possible source of oxygen for use in the fuel cell in Figure 9.15.

Figure 9.15

b) Suggest an advantage in using fuel cells rather than petrol to power cars.
   c) Write the ion–electron equation for the formation of hydrogen ions (you may wish to use the data book to help you).

3* A student set up the experiment shown in Figure 9.16.

Figure 9.16

The reaction occurring at electrode Y is:

$$Fe^{3+}(aq) + e^- \longrightarrow Fe^{2+}(aq)$$

a) Using the letters X and Y indicate the direction of electron flow.

b) Name the type of reaction taking place at electrode Y.

c) The reaction occurring at electrode X is:

$$SO_3^{2-}(aq) + H_2O(l) \longrightarrow SO_4^{2-}(aq) + 2H^+(aq) + 2e^-$$

Use the ion–electron equations for the two electrode reactions to obtain a balanced redox equation for the overall cell reaction.

4 The formula for aluminium oxide is Al_2O_3.

a) What is the charge on the aluminium ion? (You may wish to use the data book to help you.)

b) The aluminium oxide can be reduced to give aluminium metal. Write an ion–electron equation to show aluminium ions changing to aluminium atoms.

5* Figure 9.17 shows the electrolysis of molten lead(II) iodide. Which of the following describes the reaction at the positive electrode?

A I^- reduced

B I^- oxidised

C Pb^{2+} reduced

D Pb^{2+} oxidised.

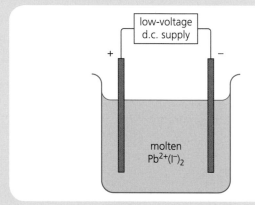

Figure 9.17

6* The electrolysis of molten sodium hydride (Figure 9.18) produces hydrogen gas.
The ion–electron equations for the reactions taking place at the electrodes are:

$$Na^+ + e^- \longrightarrow Na$$

$$2H^- \longrightarrow H_2 + 2e^-$$

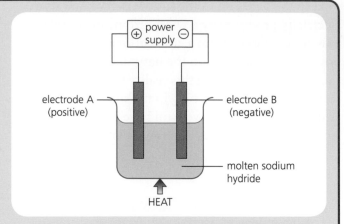

Figure 9.18

a) Use the equations to identify the electrode at which hydrogen gas is produced.

b) Combine the two ion–electron equations to give a balanced redox equation.

c) Explain how a metal, like sodium, can conduct electricity.

7* The following statements relate to four different metals, P, Q, R and S.

● In a cell, electrons flow from metal P to metal Q.

● Metal S is the only metal that reacts with cold water.

● Metal R is the only metal that can be obtained from its ore by heat alone.

The order of reactivity of the metals, starting with the *most* reactive is:

A S, P, Q, R

B R, Q, P, S

C R, P, Q, S

D S, Q, P, R.

8* Zinc reacts with copper(II) sulfate solution.

$$Zn(s) + Cu^{2+}(aq) + SO_4^{2-}(aq) \longrightarrow$$
$$Zn^{2+}(aq) + SO_4^{2-}(aq) + Cu(s)$$

a) Rewrite the equation omitting the spectator ions.

b) When iron rusts, iron atoms are converted into iron(II) ions.

i) Write the ion–electron equation for this reaction.

ii) What name is given to this type of reaction?

10 Properties of plastics

Worldwide, the plastics industry is worth trillions of pounds. Polymer chemists are working hard to produce even more bespoke plastics to fulfil needs we may not yet know that we have.

Figure 10.1 Toy building bricks

Most people would say that a toy building brick such as Lego is made of plastic (Figure 10.1). Someone with a little knowledge of chemistry, however, would say that it is made of a **polymer**. Polymers are very big molecules made from (usually) very small molecules called **monomers**. Some polymers soften or melt when they are heated; these **thermoplastic** polymers can be recycled (Figure 10.2). Polyethylene terephthalate (PET) is a thermoplastic polymer used to make such products as fizzy drink bottles, carpets and even fleeces. Like many common plastics, PET

is not biodegradable, so if it is not recycled, it will clog up landfill sites and damage the environment. This means that recycling is very important to all of us.

Figure 10.2 These fizzy drink bottles can be recycled to make new plastic products

Alkenes

In Unit 2, you learned that the unsaturated hydrocarbons known as alkenes can take part in **addition reactions**. In an addition reaction, one of the bonds in the carbon-to-carbon double bond breaks open and other atoms or groups of atoms add on to the carbon atoms in the bond. An example of an addition reaction is shown in Figure 10.3.

H$\underset{H}{\overset{H}{>}}$C=C$\underset{H}{\overset{H}{<}}$ + H—H → H—$\overset{H}{\underset{H}{C}}$—$\overset{H}{\underset{H}{C}}$—H

C_2H_4 + H_2 → C_2H_6
ethene   ethane
unsaturated   saturated

Figure 10.3 An addition reaction

Ethene molecules can also take part in addition reactions with other ethene molecules to give a very big molecule called poly(ethene); this sort of reaction is called **addition polymerisation**.

Alkenes are one of the main **feedstocks** for the plastics (polymer) industry and poly(ethene) is the most widely used plastic in the world. Several different types of poly(ethene) have now been developed and around 80 million tonnes of the polymer are sold each year. In industry, poly(ethene) is usually called polyethylene and it is sold under the trade name of polythene. Polythene was discovered by UK chemists working in the 1930s and was one of the first **synthetic polymers** to be manufactured on a large scale. Nylon (Figure 10.4) is the trade name for another mass-produced synthetic polymer, but nylon is a different type of polymer altogether: nylon is a **condensation polymer**.

Addition polymers and condensation polymers

The addition polymers we will meet in this course are made from only one type of monomer (Table 10.1). For example, poly(propene) is made from a monomer called propene and poly(phenylethene) is made from a monomer called phenylethene. The monomers used in addition polymerisation all contain the same functional group: a carbon-to-carbon double bond.

A condensation polymer, however, is made from two very different types of monomer which contain different functional groups. For example, a polyester molecule is made from dicarboxylic acid monomers and dialcohol monomers combined alternately to make a long chain. This is dealt with in more detail later in the chapter.

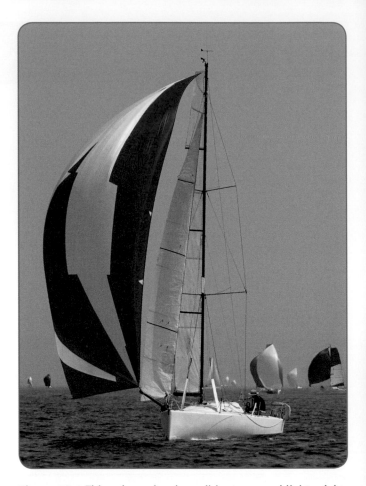

Figure 10.4 This nylon spinnaker sail is strong and lightweight

Monomer name	Polymer name
ethene	poly(ethene)
propene	poly(propene)
chloroethene	poly(chloroethene)
tetrafluoroethene	poly(tetrafluoroethene)
phenylethene	poly(phenylethene)

Table 10.1 Some common addition polymers

Identifying a polymer as an addition or condensation polymer

If we know the molecular structure of a polymer, it is easy to identify its type. If the 'backbone' chain of atoms in a polymer molecule is made entirely of carbon atoms, then it is an addition polymer. If the backbone contains oxygen (or nitrogen) atoms as well as carbon atoms, then it is a condensation polymer.

Repeating units

One thing all types of polymer molecules have in common is that they are all repetitive; the big polymer molecules are made from thousands of small monomer molecules combined together so the long chains are made of identical sections that repeat themselves over and over again. These sections are called **repeating units**.

Shapes of molecules

Making molecular models of compounds containing carbon shows that these molecules are rarely flat. Another term, often used instead of flat, is two-dimensional. If you put your hand flat on the desk, it is not two-dimensional because it is long, it is wide and it is also has a measurable depth or 'height'. A piece of paper placed on the desk is a better approximation of a two-dimensional object, but it still has a tiny, but measureable, height. If you look at a model of a water molecule, you will see that it is V-shaped and that it lies flat on the desk. Remember that an actual water molecule is around one million times smaller than the model and you will see why we describe water molecules as two-dimensional. By contrast, if you look at a methane molecule, you will see that no matter how many times you turn it over, it will never lie flat on the desk. Carbon atoms form four covalent bonds, so each carbon atom in methane is surrounded by four pairs of electrons, all negatively charged and all trying to get as far away from each other as possible. This **electronic repulsion** gives a methane molecule a particular shape: it is said to be **tetrahedral**. (Look back at Chapter 3 in Unit 1 of this book to refresh your memory on shapes of molecules.)

Addition polymers

Poly(ethene)

Poly(ethene) (old name polyethylene, trade name polythene) is made from the monomer ethene. The structural formula for ethene is as shown in Figure 10.5; however, when we are representing monomers for use in addition polymerisation, it is best to use a 'rugby goalpost' type of structure because this focuses attention on the most important part of the molecule, the double bond. The three ethene molecules shown in Figure 10.6 are drawn in this format.

Working in a group, make a model of a methane molecule. Stand it on a piece of paper, draw straight lines joining each of the three atoms then cut out the shape you have drawn. Turn the molecule over and repeat this process and keep doing it until you have a piece of paper for each side of the molecule. Use glue to attach a piece of paper to each 'face' of the model. How many faces, or sides, has the model got? Did you need to keep turning the model over to mark out the shapes? Use the internet to find out what the word 'tetrahedral' means.

Now make a model of an ethene molecule and see how the carbon-to-carbon double bond has forced the molecule to lie more or less flat on the desk with the hydrogen atoms pointing away from the double bond. Figure 10.5 gives the structural formula of ethene to help you.

Figure 10.5 Ethene

Figure 10.6 The 'rugby goalpost' structure of ethene

When the monomers are mixed in the presence of a catalyst, one of the covalent bonds in the C=C bond breaks; this means that each carbon atom is now only forming three bonds and there is a free electron on each of the carbon atoms (see Figure 10.7). These can combine with free electrons in other molecules and a chain of monomer molecules can join together.

Figure 10.7 Stages in the polymerisation of ethene

All addition polymers are formed in a similar way with thousands of monomer units linked together. The structure of each repeating unit in the chain can be obtained from the structure of the monomer molecule by breaking the double bond and leaving each carbon atom with one open bond to show it is part of a chain. For example, the repeating unit in a poly(ethene) molecule is as shown in Figure 10.8.

Figure 10.8 The repeating unit in a poly(ethene) molecule

The number of monomer units in a polythene molecule depends on the type of polythene that is being manufactured, so if we use the letter 'n' to mean any big number (more than a thousand), we can use the repeating unit to construct an equation for addition polymerisation (Figure 10.9).

Figure 10.9 Equation for addition polymerisation

Activities

Working in a group, use molecular models to make as many molecules of ethene as possible. Break one of the double bonds in each molecule and add all the molecules in the class together to form a polymer molecule. How many repeating units are in the molecule you have made? Discuss why the chain of carbon atoms is a zigzag shape rather than a straight line.

This equation can be made simpler by using molecular formulae, for example

$$nC_2H_4 \longrightarrow (C_2H_4)_n$$

(n monomer molecules) (repeating unit repeated n times in the polymer chain)

Poly(phenylethene)

Poly(phenylethene) (old name, polystyrene) is a thermoplastic polymer used to make vending cups. Blowing carbon dioxide or nitrogen gas into the polymer makes expanded polystyrene which is widely used for packaging. The molecular formula for phenylethene is C_8H_8 and the structural formula is $CH_2{=}CH{-}C_6H_5$.

The group of atoms shown as C_6H_5 is the **phenyl group**. It has a ring structure that chemists show as a circle inside a hexagon. The structure of phenylethene is shown in Figure 10.10.

Figure 10.10 The structure of phenylethene

Looking at these 'goalpost' structures, we can see that when addition polymerisation takes place, the carbon atoms from the double bond in each monomer will combine together to form the backbone chain, as in poly(ethene) (Figure 10.11). In this instance, however, every second carbon atom in the chain is bonded to a $-C_6H_5$ (phenyl) group. It is the presence of these phenyl groups which gives poly(phenylethene) its different properties.

Figure 10.11 The polymerisation of phenylethene

Using molecular formulae, the equation for the polymerisation of phenylethene becomes:

$$nC_8H_8 \longrightarrow (C_8H_8)_n$$

The information on pages 93 and 94 gives a detailed look at two of the five polymers listed in Table 10.1. Working in a group, choose one of the remaining polymers and use the information above and the internet to find answers to the following questions about your polymer:

- What are its main uses?
- Does it have an old name or a trade name?
- What is the structure of the monomer?
- Draw a section of the polymer showing four monomer units joined together.
- Work out the repeating unit and write equations for the addition polymerisation using first structural formulae and then molecular formulae.

Teach a 5-minute lesson to the class making sure they have got good notes on all the points you have researched. Make sure you are prepared to answer any questions they might ask about your polymer.

Natural polymers

There is only one naturally occurring addition polymer and that is rubber (Figure 10.12, top). Rubber is obtained from rubber trees as a white liquid (latex). All other natural polymers are condensation polymers. Most of the familiar natural polymers, such as silk (Figure 10.12, bottom), wool, hair, fur and DNA, are proteins and belong to a family called the **polyamides**. Starch

(found in bread and potatoes) and cellulose (the main constituent of wood and paper) belong to a different family of natural polymers called **polysaccharides**.

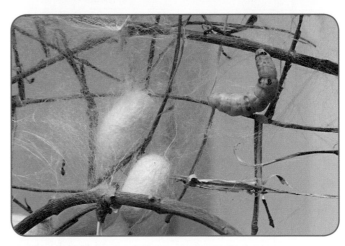

Figure 10.12 Natural polymers: rubber (top) and silk (bottom)

Research task: What is my name?

I am a naturally occurring mixture of polyesters and I have been in use for over 3000 years. I am produced by female bugs in the forests of India and Thailand, but I am edible. I have an E number and I am used as a 'wax' coating on oranges and lemons to prolong their shelf life. In the 1930s I was used all over the world as a varnish, but I have been pushed out of the market by synthetic polymers. The same is also true for records; I was used to make the original gramophone records until they started using vinyl in the 1940s. Now I am famous again! I am advertised across the internet because in 2011, a Californian company (who spent 5 years researching my properties) developed a new cosmetic treatment and I am a major part of it.

Condensation polymers

In this course, we will look at one particular type of condensation polymer: **polyesters**. In Unit 2, you learned that the group of compounds called esters are

In this experiment, you will be making rubber in the laboratory and testing its properties. (Further details can be obtained from the RSC website, **www.rsc.org**.)

Firstly, to make a rubber ball, place $10\,cm^3$ of latex solution in a small beaker. Add approximately $5\,cm^3$ of vinegar or citric acid solution and stir. Cross-links are formed in the mixture and a rubber ball is formed very quickly. Wearing vinyl protective gloves and eye protection, remove the rubber ball from the beaker and place it in a bowl of water. Squeeze the ball well to remove excess latex and then dry the ball with a cloth.

Secondly, repeat the above experiment, but add one spatula of sodium hydrogencarbonate to the beaker before adding the vinegar or citric acid solution to make foam rubber. Bubbles of carbon dioxide permeate the rubber ball as it forms.

Finally, to make an elastic band, place a pencil or the end of a test tube in the latex solution so that the end is well covered and then dip the end in a beaker of vinegar, followed by a beaker of water. The rubber can then be rolled off the end as a ring.

Working in a group, plan and design the following investigations:

a) A bounce height experiment to compare the elastic properties of the two rubber balls you have made.
b) An experiment to find out whether rubber is a thermoplastic or a thermosetting polymer.
c) An experiment to find out whether a long rubber band stretches more than a short one.

Make a list of the apparatus you will need and discuss your plans with your teacher/lecturer. Think about ways in which you could make your experiments more accurate.

Discuss among yourselves the order in which you will have to do these experiments, assign tasks to each member of the group, make samples of rubber, as described above, and test the samples you have made.

As a supplementary task, think about the following. You used an acid to make the liquid latex into a solid ball. Make a prediction about the type of chemical that might reverse this reaction and carry out an experiment to check your prediction.

---

made by the reaction of an alcohol with a carboxylic acid. The word equation for this type of reaction is:

$$alcohol + carboxylic\ acid \rightleftharpoons ester + water$$

A molecule of water is lost each time an alcohol and a carboxylic acid react together, so this type of reaction is called a condensation reaction and polymers made by this reaction are called **condensation polymers**.

You know that esters are responsible for a number of 'fruity' smells and that the ester called propyl ethanoate smells like pears. Propyl ethanoate is made from the alcohol propan-1-ol and the carboxylic acid ethanoic acid.

The equation in Figure 10.13 shows that the alcohol and acid molecules join together through their functional groups. The two monomers are then held together by a group of atoms known as an **ester link**.

In the example shown in Figure 10.13, the two molecules that have combined to form the ester each have only one functional group, so only one ester link

Figure 10.13 The formation of the ester propyl ethanoate. Note that the hydrogen atom from the alcohol and the hydroxyl group from the acid combine to form a water molecule

Figure 10.14 An ester link

can form. If we want to make a **polyester** (in other words, a long chain of alternate acid and alcohol molecules all held together by ester links), then each monomer molecule must have *two* functional groups so that the chain can continue to grow. The monomers usually used in industry to make polyesters are **diols** (alcohols with two hydroxyl groups per molecule) and **dicarboxylic acids** (acids with two carboxyl groups per molecule). The structure of these monomers can be a bit complicated, so to simplify matters, we often just show the functional groups then use the letters X or Y to represent the rest of the molecule. Using this notation,

the general reaction of a diol with a dicarboxylic acid is as shown in Figure 10.15.

In order to work backwards and obtain the structure of the monomers, given the structure of a polyester polymer, the procedure shown in Figure 10.16 can be used.

Figure 10.15 Formation of a polyester

Figure 10.16

Figure 10.17 The structural formulae for the monomers, terephthalic acid and ethane-1,2-diol

PET

This is the polymer used to make most fizzy juice bottles and it was the first commercially useful polyester to be manufactured. It is made from the monomers ethane-1,2-diol (which contains two hydroxyl groups) and terephthalic acid (which contains two carboxyl groups) (Figure 10.17). The systematic name for PET is poly(ethene terephthalate); industrially it is usually called polyethylene terephthalate, but it is best known by its trade name of 'Terylene'.

Activities

Read the section on condensation polymers and discuss in groups how to use the general reaction of a diol with a dicarboxylic acid to combine two molecules of ethane-1,2,-diol and two molecules of terephthalic acid and make a short section of a Terylene molecule. Use full structural formulae to draw the section of the Terylene molecule that would be made and circle each ester link. Identify and draw the structure of the repeating unit for this polymer.

Questions

1 a) Do you think that poly(buta-1,3-diene) is an addition polymer or a condensation polymer? Give a reason for your answer.

  b) Name the monomer used to make poly(buta-1,3-diene).

2 Figure 10.18 shows a short section of a polymer molecule.

$$-O-CH_2-CH_2-O-CH_2-CH_2-O-CH_2-CH_2-O-CH_2-CH_2-O-$$

Figure 10.18

  a) Is this an addition polymer or a condensation polymer? Give a reason for your answer.

  b) Identify the repeating unit for this polymer and draw its full structural formula.

3 Perspex, a lightweight plastic replacement for glass, is made from the monomer shown in Figure 10.19.

Figure 10.19

  Draw
  a) the repeating unit for the polymer
  b) a part of the Perspex polymer chain showing three repeating units.

4 Polymer P is a versatile plastic that can be made into a film for packaging, fibres for carpets and the 'synthetic grass' used for some hockey and football pitches. Its structure is shown in Figure 10.20.

Figure 10.20

The repeating unit in the polymer is shown in Figure 10.21.

Figure 10.21

The monomer M from which the polymer is made is shown in Figure 10.22.

Figure 10.22

  a) Name the monomer M and the polymer P.
  b) To which homologous series does M belong?

5 When superglue sets, a polymer is formed. The polymer has the structure shown is Figure 10.23.

Figure 10.23

Draw the structural formula for the monomer in superglue.

6 Polyvinylidene chloride is an addition polymer. It is added to carpet fabrics to reduce flammability. It is made from the monomer vinylidene chloride. Part of the polymer molecule is shown in Figure 10.24.

Figure 10.24

  a) What is meant by an *addition* polymer?
  b) Draw the full structural formula of the vinylidene chloride monomer.

Novel materials

A **novel material** is a new material; one that does not resemble anything that was previously known or used. Since the nineteenth century, scientists have discovered thousands of novel materials, and these have changed our lives in so many ways, primarily by turning us all into consumers. Some of these new materials exhibit 'smart' behaviour. A smart material is one that changes shape in response to changes in its environment. The change in shape might be on the macroscopic level (so we can see it) or on the microscopic level so we cannot see it, but we can see its effects. Hydrogels are smart materials that have found a lot of uses in daily life because they are polymers that absorb water.

Activities

Working in a group, discuss the order in which you are going to tackle these activities. Assign tasks to each group member. If time is short, it might be more effective to split the activities between you; the students who take on Activity 1 could then report back to the main group. List the apparatus you need before you start the experimental work, make a careful note of everything you observe then wash and dry the apparatus when the experiment is over.

1 Hair gel

Use a teaspoon to put a spoonful of the same brand of hair gel onto two separate petri dishes. Sprinkle two spatulas of salt over one sample of hair gel and stir the mixture. Using a clean spatula, sprinkle two spatulas of sugar over the other sample and stir the mixture. Stir each sample occasionally and watch what happens over the next 5 minutes. While you are observing, think about the following questions:

a) Which elements are present in common salt? What is the chemical name of common salt? What type of bonding do you expect it to have?
b) Sugar (sucrose) is made from the elements carbon, hydrogen and oxygen. What type of bonding do you expect to find in sugar?
c) What type of particles might be responsible for the effect salt has on hair gel?

2 Disposable nappies

Cut the centre (the thick part that is designed to absorb urine) out of an ultra-absorbent disposable nappy and put it into a clean, dry, weighed ice-cream tub. Carefully pull the central part of the nappy apart and shake it gently to get the small white grains of hydrogel into the bottom of the tub. Be careful not to get too much fluff and dust mixed with the grains. When you think you have extracted all the hydrogel, pick out any fluff from the tub and discard it with the rest of the nappy. Weigh the tub again and calculate the mass of hydrogel. Add about 100 cm³ of distilled water (or tap water) and stir with a glass rod. Keep adding water until you are at the limit of the polymer's absorption then weigh the tub again. Calculate the mass of water absorbed and find the ratio of mass of hydrogel: mass of water absorbed.

Have a group discussion on the results of Activity 1 and predict the effect that adding some salt or sugar might have had on the material in the tub.

Test your predictions, record the results and answer the following questions.

1 Why is it better to use distilled water, rather than tap water, in this experiment?
2 Will the nappy absorb as much urine as it absorbed distilled water? Give a reason for your answer.

Figure 10.25

Figure 10.26

Hydrogels as smart materials

The first hydrogels were used in the 1950s to develop early contact lenses and many different types have now been developed. In simple terms, they are all insoluble substances that form a tangled, three-dimensional network of fibres. When a gel is in contact with water, functional groups along the fibres (often carboxyl groups) ionise and the negative charges repel each other, forcing the fibres to become less tangled and to expand. Water is a covalent substance, but water molecules carry small, electrostatic charges and they are attracted into the porous network of fibres. If an ionic compound (such as NaCl) is added to the mixture, the Na^+ ions partially neutralise the negative sites on the fibres and the expanded polymer collapses back to its original shape, expelling the water molecules. Hydrogels are smart materials because their molecules change shape in response to the concentration of ions in the surrounding solution.

Changing colours

Smart polymers have been developed that change colour depending on the temperature. These can have serious uses, for example, in the manufacture of a type of thermometer that can be used to find out whether a child has a temperature by simply laying it on their forehead, and novelty uses, for example, when they are incorporated into ceramic mugs which suddenly change colour when a hot drink is poured in.

A soluble polymer

Most polymers are insoluble in water, but polyvinyl acetate (PVAc) dissolves in water because the acetate groups are electrostatically attracted to water molecules.

Slime

To make slime, add 25 cm³ of PVAc solution and five drops of food colouring to a plastic beaker, followed by borax solution, added 1 cm³ at a time. Stir the mixture vigorously after each addition of borax (wooden lollipop sticks are useful stirrers) until the slime wraps around the stirrer and can be lifted out of the beaker and moulded into a sphere. The volume of borax needed to achieve the required consistency should be noted.

Activities

Make three samples of different coloured slime.

Sample 1: made as above.
Sample 2: use 25 cm³ PVAc solution + half the volume of borax solution used in sample 1.
Sample 3: use 25 cm³ PVAc solution + one and a half times the volume of borax solution used in sample 1.

Examine your samples carefully and write down two properties of slime. Leave a ball of slime on the desk for 5 minutes and watch how the solid behaves.

Plan and carry out experiments to find out which sample will stretch the most without breaking and which sample flows the fastest.

Note: Polyvinyl acetate (PVAc) is found in products called wood glue, white glue or simply PVA glue, and should not be confused with polyvinyl alcohol which is also referred to as PVA. Borax causes the PVAc molecules to cross-link and form the unusual material called 'slime'.

A polymer that conducts electricity

Poly(ethyne) is a hydrocarbon, so it is a covalent compound that should be an insulator. It is made by polymerising ethyne ($HC{\equiv}CH$), and a short section of the polymer can be drawn as:

$$-CH=CH-CH=CH-CH=CH-$$

The backbone contains alternating carbon-to-carbon double and single bonds. However, the extra electrons in the double bonds are not confined to one place – they are delocalised along the chain. Consequently, this material is a good conductor of electricity and, when suitably processed, its electrical conductivity compares well with the best metallic conductors like copper and silver.

Smart metals

An **alloy** is a mixture of a metal and one or more other elements, usually other metals. Alloys have been used for thousands of years. They are designed to have particular properties so they can do a specific job. For example, stainless steel (iron + carbon + chromium) does not rust.

One recently invented group of alloys share a quite spectacular property. These are known as **shape memory alloys** (**SMAs**). These metal mixtures can 'remember' their original shape and when they are heated above a certain temperature (the transition temperature), they will return to that shape (Figure 10.27), no matter how much they have been bent or folded. They do this because atoms in a piece of metal are arranged in a particular way (we say metals have a crystal structure) and at their transition temperatures, SMAs change from one type of crystal structure to another. When they change their structure, they change their shape. The first SMAs were nickel–titanium alloys but there are now many different SMAs with widely varying transition temperatures and uses.

Stents are devices used to hold open clogged arteries, usually around the heart. SMA stents are very thin pieces of metal which are inserted into a big artery in the thigh and pushed all the way up the artery to the correct position near the heart. They have a transition temperature close to body temperature, so when they are in position, the shape changes to a coil or cylinder which holds the artery open. The surgeon has to work pretty fast!

Many building that are unoccupied for periods of time, for example overnight or at weekends, use automatic sprinkler systems to attempt to put out fires that may start accidentally. Some systems use pieces of SMA which, as the fire starts and the temperature rises, change shape and trigger the sprinklers.

GPS and mobile phone satellites have aerials with a big surface area. These would be damaged during launch so they are folded on SMA frames which open out when the satellite is in orbit and exposure to the Sun's rays has made them reach their transition temperatures.

Figure 10.27 A thin wire of a shape memory alloy

Checklist for Revision

- I understand the difference between addition and condensation polymerisation.
- I can represent the structures of monomers and polymers using structural formulae.
- I can name some natural polymers.

End-of-chapter questions

1* PVC is a very useful plastic. It can be used to make window frames and to make the insulating cover for electrical cables. A short section of the polymer is shown in Figure 10.28.

Figure 10.28

a) Draw the repeating unit present in the polymer.
b) Draw the full structural formula of the monomer used to produce the polymer PVC.
c) PVC is not biodegradable.
   i) What is meant by the term biodegradable?
   ii) Why have biodegradable polymers been developed?

2* Polymers are used in the manufacture of glass windscreens (Figure 10.29). This makes the windscreen less likely to shatter into fragments.

glass — polyester film
— PVB polymer
— polyester film

Figure 10.29

a) Name the type of polymerisation reaction used to make polyesters.
b) The PVB polymer is made from a monomer with the structure shown in Figure 10.30.
Draw a section of the polymer showing three monomer units linked together.

H   H
| |
C = C
| |
H CH_2CHO

Figure 10.30

3* Poly(ethenyl ethanoate) is an addition polymer. Part of its structure is shown in Figure 10.31.

H $OCOCH_3$ H $OCOCH_3$ H $OCOCH_3$
| | | | | |
—C—   —C—   —C—   —C—   —C—   —C—
| | | | | |
H   H   H   H   H   H

Figure 10.31

a) Draw the structural formula for the monomer used to make this polymer.
b) Poly(ethenyl ethanoate) is used to make another polymer and methyl ethanoate is a by-product of that reaction.
   i) What type of compound is methyl ethanoate?
   ii) Draw the structure of methyl ethanoate.

4 A non-stick frying pan is made of metal coated with poly(tetrafluoroethene). A section of the polymer is shown in Figure 10.32.

F   F   F   F   F   F
| | | | | |
—C—C—C—C—C—C—
| | | | | |
F   F   F   F   F   F

Figure 10.32

a) What type of polymer is poly(tetrafluoroethene)?
b) Draw the structure of the monomer from which this polymer is made.

5 One hundred years ago people only used natural polymers, but now we use a lot of smart materials because they have properties that are useful in everyday life.
a) What is a smart material?
b) A hydrogel is a smart material; give one common use for hydrogels.
c) Name two natural polymers.

11 Fertilisers

Food security

Food security is a term frequently used in news programmes. The UN Food and Agricultural Organization defines food security as, 'acccss to sufficient, safe and nutritious food to meet [a person's] dietary needs and food preferences for an active and healthy life'. The issues associated with food security are covered in depth in the National 5 Biology course, so here we will look at how twentieth-century chemists developed the artificial fertilisers that help feed the world. In the National 4 Chemistry course you may have learned about the impact of synthetic and natural fertilisers on the world's food production, but to recap on the importance of fertiliser manufacture, we need a few facts and figures.

World population

A look at the US census site (given below) will show you an estimate of the number of people who presently share this planet.

www.census.gov/main/www/popclock.html

The 'Green Revolution' of the twentieth century increased food supplies and population growth followed. Experts agree that the world's population is still increasing (although the rate of growth is now decreasing) and the statistics indicate that there will be over 9 billion of us by 2050. Artificial fertilisers play a major role in feeding hungry people, but their manufacture depends on the availability of fossil fuels, and fossil fuels are a finite resource. Water is also becoming a vital resource (Figure 11.1), as patterns of rainfall change across the planet, so new methods of irrigation and water storage need to be developed.

Elements essential for plant growth

Healthy crops provide more food and to be healthy, plants need about 16 different chemicals, the three most important ones being nitrogen, phosphorus and potassium (NPK). When crops grown locally were eaten locally, these elements could be returned to the soil by using natural fertilisers. As more people moved into towns, crops were harvested and, to feed the townspeople, they were taken miles from where they had been grown. As a result, natural fertilisers were no longer enough to nourish the soil. Nitrogen is one of the most important elements needed by plants and, for many years, ships ferried guano (bird droppings) from Peru to Europe to use as fertiliser because guano contains a lot of nitrogen, as well as phosphorus and other useful chemicals. If you go to the Isle of May in summer, when the sea birds are nesting, you can see guano on the cliffs (Figure 11.2). In this country, it does not stay on the cliffs for long because it is soluble in water, another property that a fertiliser must possess, and so gets washed away by the rain. In the dry conditions of South America, huge deposits built up and, during the nineteenth century, they were dug out and used to fertilise the fields of Europe.

Figure 11.1 The Sahara desert

Figure 11.2 Sea cliffs white with guano

Nitrogen fixation

Farmers have known for possibly thousands of years that some leguminous plants, such as peas, beans and clover, can be used as 'green compost'. In other words, they can be grown and then ploughed straight back into the soil to improve its fertility. This is because these plants contain colonies of bacteria which trap (or fix) nitrogen from the air and use it for growth. Although it makes up 80% of our atmosphere, nitrogen is an unreactive gas; it is difficult to fix. A brilliant chemist called Fritz Haber started working on this problem in 1900. By 1908 he had successfully fixed atmospheric nitrogen in the laboratory to make the compound called ammonia (NH_3) and the stage was set for one of the biggest developments of the twentieth century: intensive agriculture.

Formation of ammonia from its elements

There is a more or less limitless supply of nitrogen in our atmosphere and hydrogen can be obtained from natural gas (methane), so the elements used to make ammonia are relatively cheap and easily available. The compound itself was first isolated in the eighteenth century, but back then it was not prepared from its elements, and the method of preparation made it too expensive to be produced industrially. Haber's work inspired chemists and engineers to scale the process up from a laboratory reaction to a profitable chemical industry. To make money, the industry needed a fast reaction that gave a high yield of ammonia at the lowest possible cost, but scientists faced several problems.

1 The reaction between nitrogen (N_2) and hydrogen (H_2) to make ammonia (NH_3) is **reversible**:

$$N_2(g) + 3H_2(g) \rightleftharpoons 2NH_3(g)$$
$$\text{(reactants)} \qquad \text{(product)}$$

You know from your work on esters that a double arrow, as given in this equation, indicates that a reaction is reversible. Reversible means that nitrogen and hydrogen are combining to form ammonia, but at the same time, ammonia is decomposing to form nitrogen and hydrogen.

When N_2 and H_2 are first mixed and start to react, the concentration of the elements is high and the concentration of the compound (NH_3) is low (because very little of it has yet formed). As you know from Unit 1 of this course, a high concentration of reactants means a fast reaction, and vice versa, so initially the reaction from left to right is fast and the reaction from right to left is slow. As more of the N_2 and H_2 react and get used up, the reaction from left to right (usually called the **forward reaction**) slows down and, as more of the NH_3 is formed, the reaction from right to left (usually called the **reverse reaction**) speeds up. Eventually, there comes a point at which ammonia is breaking down at the same rate as it is being formed. This means that it is impossible to change all the nitrogen and hydrogen into ammonia. In fact, this reaction gives a low yield of ammonia.

2 The reaction is also slow at room temperature. Both forward and reverse reactions can be speeded up by increasing the temperature, but ammonia decomposes much faster at higher temperatures, so increasing the temperature decreases the ammonia yield still further.

3 The reactants (N_2 and H_2) are gases, so the molecules are far apart, yet the molecules must collide if they are to react. Exerting pressure on the system 'squeezes' the molecules closer together and increases the yield of ammonia, but running an industrial process at high pressures is costly and the higher the pressure, the greater the production costs.

It took 5 years for Haber, working with a research team led by the engineer Karl Bosch, to develop the first industrial-scale application of the Haber process (Figure 11.3). The process was carried out in a sealed system kept at high pressure. A catalyst (powdered iron) was used to get a fast reaction at a relatively low temperature. The difference in boiling points between the three gases, N_2, H_2 and NH_3, allowed the engineers to condense ammonia and run it off as a liquid, leaving the unreacted gases (nitrogen and hydrogen) to be recycled to avoid wasting reactants.

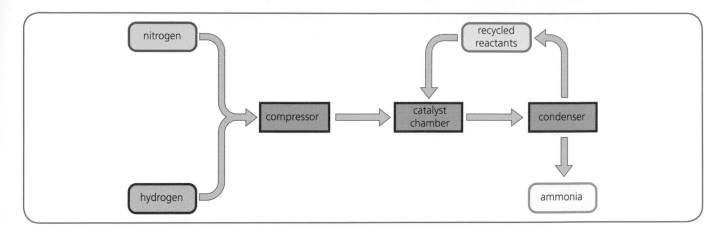

Figure 11.3 In the Haber process, nitrogen and hydrogen are combined to make ammonia

Ammonia

Ammonia is an important chemical because it is a feedstock for the manufacture of nitric acid (Figure 11.4) and for the fertiliser industry.

Properties of ammonia

Ammonia gas consists of NH_3 molecules. It is soluble in water, because the molecules react with water to form ammonium ions, $NH_4^+(aq)$, and hydroxide ions, $OH^-(aq)$. This is a reversible reaction.

Figure 11.4 An industrial nitric acid production plant

Activities

Industrial chemists in different countries use different conditions for the production of ammonia, depending on the plant they are operating. Go to the website **www.freezeray.com/chemistry.htm** and use the animation of the Haber process to find out how the yield of ammonia changes as the conditions of temperature and pressure change.

Construct a graph with percentage yield of ammonia on the vertical axis and pressure in atmospheres (atm) on the horizontal axis. Set the temperature at 200 °C and the pressure at 100 atm, discharge the ammonia as soon as 'equilibrium' is reached and note the yield. Keep the temperature constant, increase the pressure to 200 atm and note the yield again. Note the yield of ammonia at 400 atm, 600 atm, 800 atm and

1000 atm, then plot the readings on the graph. Draw a line of best fit and label the line 200 °C.

Choose two different temperatures and repeat the process for each temperature. You now have three lines on your graph. Decide whether these three give you enough information about yields in the Haber process. Should you draw a fourth?

Remembering that high pressures generate high running costs, use your graph to estimate the most cost-effective conditions for the manufacture of ammonia. Use the internet to check your prediction.

(The website quoted in this activity can also be found via the Education Scotland website in the Chemistry Support document: *Agrochemicals, Advice and Guidance for Practitioners*, page 19.)

Be careful not to get ammonia molecules, NH_3, confused with ammonium ions, NH_4^+.

$$NH_3(g) + H_2O(l) \rightleftharpoons NH_4^+(aq) + OH^-(aq)$$

Ammonia solution (sometimes called ammonium hydroxide solution) has a pH above 7 and turns universal indicator solution blue. Ammonia is the only alkaline gas you will meet in this course – in other words, the only gas that turns damp pH paper blue. It does this because it dissolves in the water in the damp paper to form hydroxide ions.

The high solubility of ammonia can be shown by the fountain experiment (Figure 11.5).

Figure 11.5 All the ammonia gas in the flask can dissolve in just a few drops of water. When this happens, the water from the beaker rushes into the flask to fill the space, creating a fountain effect as soon as the clamp is released

Care should be taken when using ammonia gas. It is the main active ingredient of smelling salts and it irritates the mucous membranes of the nose and lungs.

Laboratory preparation of ammonia

Ammonia can be made by heating any solid ammonium compound with a solid alkali, such as soda lime or calcium hydroxide. If solutions are used instead of solids, the gas formed immediately dissolves in the water, rather than being released.

calcium hydroxide + ammonium sulfate $\longrightarrow$
  calcium sulfate + ammonia + water

Questions

1 Write a balanced chemical equation for the reaction between calcium hydroxide and ammonium sulfate.

Activities

Plan and design an experiment to prepare and test some ammonia. Draw a diagram (in pencil) of the apparatus you will need and write down how you intend to test for the presence of ammonia gas. List the chemicals you will use, then risk assess each chemical and each procedure. Think about how you will heat the mixture of solids safely and discuss your plans with your teacher/lecturer.

If you are allowed to continue, set up the apparatus then prepare and test the gas. While the apparatus is cooling, record your observations and write down two ways in which you could have improved the experiment.

How did you know that you had made ammonia gas? How could you have collected a dry sample of the gas, if you had wanted to? (Hint: look up the density of ammonia gas.)

Check that the apparatus is cool, then dismantle it, wash and dry the glassware and return it to the correct place.

Reactions of ammonia

Reaction of ammonia with acid

In the same way that ammonia reacts with water to form ammonium ions and hydroxide ions, ammonia reacts with acids to form ammonium salts. For example, ammonia reacts with nitric acid to form the salt ammonium nitrate:

$$NH_3(g) + HNO_3(aq) \longrightarrow NH_4^+(aq) + NO_3^-(aq)$$
ammonia + nitric acid $\longrightarrow$ ammonium nitrate

Plants can only obtain the essential elements they need for healthy growth if compounds containing those elements are dissolved in soil water, the water that plants absorb through their roots. This means that fertilisers have to be soluble in water. A look at the solubility table in your data booklet will tell you that all ammonium compounds are soluble in water, so as ammonium compounds also contain nitrogen, they are often used as fertilisers.

Oxidation of ammonia

Ammonia can, in the presence of a catalyst, be oxidised, and this is the first step in the manufacture of nitric acid.

Many common fertilisers are salts of nitric acid, so we shall first look at the manufacture of nitric acid before we look at the huge agrochemical industry involved in the manufacture of fertilisers.

Oxides of nitrogen

As you learned in Unit 1 of this course, soluble non-metal oxides form acids when they dissolve in water. When nitrogen dioxide (NO_2) dissolves in water, nitric acid is formed.

Nitrogen exists in the atmosphere as diatomic molecules with the two atoms in the N_2 molecule held together by a triple covalent bond. It takes a lot of energy to break these three bonds (to separate the atoms), so atmospheric nitrogen is unreactive. If, however, enough energy is available to split the N_2 molecules, the free atoms released are very reactive. Flashes of lightning (Figure 11.6) and the spark plugs in cars provide enough electrical energy to split nitrogen molecules causing nitrogen dioxide to form. Other oxides of nitrogen also exist. Ultra-violet light (sunlight) can split up NO_2 to give NO (nitrogen monoxide) which can then recombine with oxygen to form NO_2. Nitrogen

dioxide can dimerise to make dinitrogen tetroxide, and then there is N_2O which used to be used by dentists as an anaesthetic and is commonly called laughing gas.

It requires a huge amount of electrical energy to make NO_2 directly from N_2, so it is too expensive to make nitric acid commercially this way. In 1902, Wilhelm Ostwald developed (and with great forethought, patented) a method of converting ammonia into nitric acid using platinum as a catalyst. As soon as ammonia started to be mass produced, the feedstock became cheap and easily available, and the Ostwald process rapidly became a mainstay of the chemical industry. At present, some 60 million tonnes of nitric acid are produced annually, of which 85% is used to make fertilisers.

Questions

2 You know the meaning of the word 'polymerise', so deduce the meaning of the word 'dimerise'.
3 Write the formula for the compound called dinitrogen tetroxide.
4 What is the correct name for the compound N_2O?
5 In the presence of ultra-violet light, one molecule of nitrogen dioxide and one molecule of oxygen react to produce one molecule of nitrogen monoxide and one molecule of another substance. Write the equation for this reaction, and by balancing the atoms on both sides, find the formula for the unknown substance. Use the internet to find its name and to find out about the role it plays in the atmosphere.

Oxidising ammonia in the laboratory

Your teacher/lecturer may demonstrate the experiment shown in Figure 11.7.

The gas mixture used in the Ostwald process is approximately 10% ammonia and 90% air, so a stream of air needs to be pumped into the flask of ammonia to keep this reaction going. Heat is needed to start the reaction, so the piece of platinum wire needs to be red hot before it is placed in the flask. Once the reaction starts, the platinum continues to glow red because it is an exothermic reaction and the heat from the reaction

Figure 11.6 Flashes of lightning help to add nitrogen to the soil

Figure 11.7 The laboratory oxidation of ammonia

keeps the wire hot. The ammonia is oxidised to form nitrogen monoxide and water.

As the nitrogen monoxide reaches the mouth of the flask, it combines with more oxygen to form nitrogen dioxide, which is a brown gas. If this brown gas is tested with damp pH paper, the paper will turn red as the NO_2 reacts to form nitric acid.

The Ostwald process

In the Ostwald process (Figure 11.8), the mixture of gases (ammonia and air) is heated then passed through the layers of platinum gauze that act as a catalyst.

The gauze does not need to be heated separately: the oxidation of ammonia is exothermic, so the temperature stays high in the catalyst chamber and the reaction occurs at around 800 °C. The oxidation of ammonia is one of the most efficient catalytic reactions, with conversion of up to 98%.

ammonia(g) + oxygen(g) $\longrightarrow$
$\qquad$ nitrogen monoxide(g) + water(g)

The nitrogen monoxide formed reacts with oxygen to give nitrogen dioxide.

nitrogen monoxide(g) + oxygen(g) $\longrightarrow$
$\qquad$ nitrogen dioxide(g)

The brown gas is then cooled, mixed with more air and passed through a flow of water to make nitric acid.

nitrogen dioxide(g) + oxygen(g) + water(l) $\longrightarrow$
$\qquad$ nitric acid(aq)

Questions

6 Why is the platinum catalyst used in the form of a gauze?
7 Write and balance a chemical equation for each of the word equations in this section.

Uses of nitric acid

Nitric acid is unusual. It played a part in saving the lives of millions of hungry people across the planet through its use in the manufacture of fertilisers and it played a part in killing millions of people through its use in the manufacture of explosives. Nitric acid proves the point that there is no such thing as a dangerous chemical, there are only dangerous people.

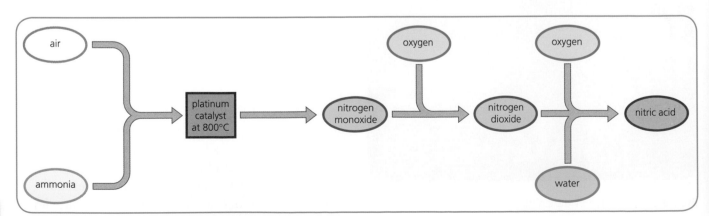

Figure 11.8 An outline of the Ostwald process

Fertilisers

If plants are grown in the same field year after year, the soil loses its nutrients, so fertilisers are used to replace them (Table 11.1). The use of fertilisers allows farmers to increase the amount of crops they grow, and that means more food can be produced and intensive agriculture becomes possible. Farmers mainly use artificial fertilisers to supply nitrogen, phosphorus and potassium (Figure 11.9). Plants take in these nutrients with the water they absorb through their roots, so fertilisers must be soluble. The solubility table in your data booklet will tell you that some types of compounds are more soluble than others, which is why compounds commonly used in fertilisers include ammonium phosphate, $(NH_4)_3PO_4$, and potassium sulfate, K_2SO_4.

Elements needed by plants	Compounds present in fertiliser
potassium	potassium salts
phosphorus	phosphates
nitrogen	nitrates and ammonium salts

Table 11.1

Figure 11.9 A bag of NPK fertiliser containing the nutrients nitrogen (N), phosphorus (P) and potassium (K)

Ammonia and nitric acid are generally produced in the same integrated chemical plant and are then often reacted together, on site, to produce ammonium nitrate, a very soluble salt with a high nitrogen content. The big fertiliser factories are mainly located in the former USSR, the USA and in central Europe, but large quantities of potassium chloride and mineral phosphates are being produced from the Dead Sea (Figure 11.10). These were sold as raw materials, but are increasingly being converted into compound fertilisers at the site of their extraction.

Figure 11.10 Salt pans beside the Dead Sea

Chemical companies now 'tailor' their fertilisers for different crops and different countries. For example, cereals need more nitrogen than potatoes, but both need phosphorus, potassium and other minerals, and farmers in a country with a heavy rainfall might use a less-soluble compound (like urea) so that the expensive fertiliser will not be washed out of the soil before plants can absorb it.

Compounds used as fertilisers can be produced in the laboratory by neutralisation reactions similar to those you looked at in Unit 1 of this course.

Questions

8 Copy and complete Table 11.2. The first row has been done for you.

Acid	Formula of acid	Ions present in aqueous solution
phosphoric	H_3PO_4	$3H^+(aq)$ and $PO_4^{3-}(aq)$
sulfuric		
nitric		
hydrochloric		

Table 11.2

9 Phosphoric acid, H_3PO_4, is used to make phosphate fertilisers and can be prepared by the reaction of calcium phosphate with sulfuric acid. Given that calcium sulfate is the only other product, write a balanced chemical equation for the reaction.

10 Potassium chloride can be made by neutralising hydrochloric acid with potassium hydroxide.
   a) Write and balance a chemical equation for this reaction.
   b) Name the spectator ions in the reaction.
   c) Write an equation for the reaction omitting the spectator ions.

11 Write and balance chemical equations for each of the following word equations.
   a) potassium hydroxide + nitric acid →
                              potassium nitrate + water
   b) ammonia + hydrochloric acid →
                              ammonium chloride
   c) ammonia + phosphoric acid →
                              ammonium phosphate

12 Complete the following word equations then write a balanced chemical equation for each of them.
   a) potassium hydroxide + sulfuric acid →
   b) sodium hydroxide + nitric acid →
   c) potassium hydroxide + phosphoric acid →
   d) ammonia + sulfuric acid →

The percentage mass composition of fertilisers

When fertilisers are prepared for sale, different chemical compounds are mixed in different proportions for different uses, so it is important to know the exact quantity of each element in fertilisers. To establish this we need to know how to calculate the percentage mass composition of a compound. First, we will revise some earlier work.

Formula mass

The formula mass of a compound is found by adding together the relative atomic mass of each atom in the formula. Values for the relative atomic masses of some elements can be found in the data booklet.

For example, if the formula is NaOH, the formula mass is $23 + 16 + 1 = 40$.

If more than one atom of an element is present in the formula, the relative atomic mass of the element is multiplied by the appropriate number.

For example, if the formula is Na_2CO_3, the formula mass is $(2 \times 23) + 12 + (3 \times 16) = 106$.

If the formula of the compound contains ions with more than one type of atom (usually called group ions), the mass of every atom inside the bracket is multiplied by the number outside the bracket.

For example, if the formula is $(NH_4)_2SO_4$, the formula mass is $(2 \times 14) + (8 \times 1) + 32 + (4 \times 16) = 132$.

Remember that in order to avoid unforced arithmetic errors, it is often safer to calculate the mass of the atoms inside the bracket then multiply that mass by the number outside the bracket.

For example, if the formula is $(NH_4)_2SO_4$, the formula mass is $[14 + (4 \times 1)] \times 2 + 32 + (4 \times 16) = 36 + 32 + 64 = 132$.

Percentage composition

Once we have calculated the formula mass of a compound, we can find the percentage by mass of any element in the compound by using the following equation:

$$\text{percentage mass of a given element in a compound} = \frac{\text{mass of element in the formula}}{\text{formula mass}} \times 100$$

Worked examples

1 $\dfrac{\text{percentage of sodium}}{\text{in sodium hydroxide}} = \dfrac{\text{mass of sodium in formula}}{\text{formula mass}} \times 100$

$= \dfrac{23}{40} \times 100 = 57.5\%$

2 $\dfrac{\text{percentage of carbon in}}{\text{sodium carbonate}} = \dfrac{12}{106} \times 100 = 11.32/11.3\%$

3 $\dfrac{\text{percentage of oxygen in}}{\text{ammonium sulfate}} = \dfrac{64}{132} \times 100 = 48.48/48.5\%$

Checklist for Revision

● I know the best operating conditions for the Haber process to produce ammonia and can explain why they are applied.
● I appreciate the need for the commercial production of nitrate fertilisers and can explain how they can be manufactured from nitric acid.
● I can calculate the percentage mass composition of fertilisers.

End-of-chapter questions

1 Calculate the percentage by mass of nitrogen in each of the following compounds which are used as fertilisers:
   a) sodium nitrate
   b) calcium nitrate
   c) ammonium sulfate
   d) ammonium nitrate.

2 Potassium is a major plant nutrient. It promotes the rate of plant growth. Potassium is usually added to soils in the form of the salt potassium chloride. Calculate the percentage by mass of potassium in potassium chloride.

3 Ammonia gas is converted into various solid compounds that are used as fertilisers. In one such reaction, the compound urea is formed. Molecules of urea have the structure shown in Figure 11.11.

Figure 11.11

   a) i) Calculate the formula mass of urea.
   ii) Calculate the percentage by mass of nitrogen in urea.
   b) Ammonia and carbon dioxide react together, using high pressure and temperature, to give urea and water. Write a balanced equation for this reaction.

4 The chemical industry produces large quantities of nitric acid by the catalytic oxidation of ammonia (Figure 11.12).

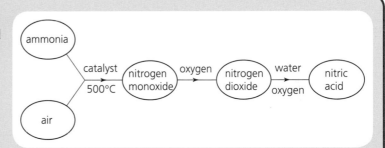

Figure 11.12

   a) State the name of this industrial process.
   b) Name the catalyst used to speed up the reaction.
   c) The nitric acid can be used to make potassium nitrate fertiliser. Write the formula for potassium nitrate.

5 As the world population increases, the demand for food grows. In order to meet this demand, farmers are using more and more synthetic fertilisers to improve crop yields. One of these synthetic fertilisers is Nitram. The flow diagram in Figure 11.13 shows how Nitram can be made industrially.
   a) i) Name reactant A.
   ii) Name industrial process X.

Figure 11.13

b) In process X, the percentage conversion of nitrogen to ammonia decreases as the temperature increases. Why, then, is process X carried out at the relatively high temperature of 450 °C?

c) What is the chemical name for Nitram?

d) Nitram is very soluble in water and this allows essential elements to be taken in by the roots of crop plants very quickly. Suggest why Nitram's high solubility can also be a disadvantage in its use as a fertiliser.

6* Plants use a compound called chlorophyll to obtain energy from the Sun, and nitrogen is one of the elements needed to make chlorophyll. Most plants get the nitrogen they need from compounds of ammonia and nitrates in the soil.

a) Name a substance that would produce ammonia if it was heated with ammonium carbonate.

b) The Ostwald process is used to produce nitric acid. The first stage in this process is the oxidation of ammonia.

   i) Which metal is used as a catalyst in the Ostwald process?

   ii) Why is it not necessary to keep heating the catalyst once the reaction has started?

7 Diammonium hydrogenphosphate is a major fertiliser made from ammonia and phosphoric acid.

$$2NH_3 + H_3PO_4 \rightarrow (NH_4)_2HPO_4$$

Calculate the percentage, by mass, of nitrogen and phosphorus in this fertiliser.

8 Phosphorus is an important plant nutrient as it regulates leaf development and size. Phosphorus is found as calcium phosphate, $Ca_3(PO_4)_2$, but this is converted into calcium dihydrogenphosphate, $Ca(H_2PO_4)_2$, which is then used as a fertiliser.

a) Calculate the percentage by mass of phosphorus in

   i) calcium phosphate

   ii) calcium dihydrogenphosphate.

b) Calcium phosphate is not used directly as a fertiliser, but calcium dihydrogenphosphate is. Suggest a reason for both of these facts. (The table of solubilities in the data booklet may help you to answer this question.)

12 Nuclear chemistry

The particle zoo

Figure 12.1 Professor Brian Cox

The range of sub-atomic particles found in the Universe is sometimes called the particle zoo. You learned about protons, neutrons and electrons in Unit 1 of this course, but how about glueons, quarks, mesons, muons and neutrinos? If you follow Robin Ince or Brian Cox (Figure 12.1) on Twitter then you might have heard of some of these particles. You may also have heard of the Higgs-Boson particle, which scientists at CERN in Switzerland believe they may have found. This particle is thought to convert energy into matter. At present that fact is uncertain, but it is certain that we, on this planet, can convert matter from the nucleus of an atom into energy; this is called **nuclear energy**.

Nuclear radiation

We live our lives in an invisible ocean of electromagnetic radiation. Radios, televisions, mobile phones (Figure 12.3), microwave ovens and WiFi networks all use electromagnetic radiation of varying wavelengths. Nuclear radiation has a different source; it is emitted from the nucleus of a radioactive isotope and it may be in the form of particles or waves.

Activities

Marie Curie and Albert Einstein were two of the scientists who revolutionised nuclear science and each faced significant barriers as they attempted to pursue their scientific work. As you work through this section on nuclear chemistry, create a timeline following the work of these two Nobel Prize winners and the development of the understanding of nuclear radiation and its applications through the twentieth and twenty-first centuries.

Figure 12.2 At the Solvay Conference held in 1927, many of the most famous scientists of the day gathered to discuss quantum theory

Figure 12.3 This smartphone uses electromagnetic radiation

Isotopes and nuclide notation

An atom of an element with a particular mass number is called an isotope or **nuclide** (nuclide is the term mainly used in the nuclear industry). The existence of isotopes means that scientists use nuclide notation to show the number of sub-atomic particles in an atom or ion. As you learned in Unit 1, isotopes are atoms that have the same atomic number (so they have the same number of protons) but they have different mass numbers (so they have different numbers of neutrons). Almost all elements have more than one naturally occurring isotope; the number varies from two up to about ten.

Hydrogen has three isotopes. The symbols for the three types of hydrogen atom are as shown in Figure 12.4.

Figure 12.4 The three isotopes of hydrogen

Rather than use superscripts and symbols to give the mass number of an isotope, many textbooks use names and numbers. For example, $^{241}_{95}Am$ can be written as americium-241. The atomic number is left out because the data booklet will tell you that if the element is americium, the atomic number must be 95. The mass number is stated because it changes as the isotope changes.

Radioisotopes

As we move through the Periodic Table, the atomic number of the elements increases and atoms get bigger and heavier with the number of neutrons in the nucleus increasing more rapidly than the number of protons. As the ratio of neutrons to protons increases, the nuclei can become unstable. An unstable nucleus (or atom) is said to be **radioactive**. These radioactive nuclei contain a lot of energy and they tend to rearrange themselves spontaneously to reduce the size of the nucleus and emit some energy. This process is called **nuclear decay** and isotopes that decay are called **radioisotopes**. There are a relatively small number of naturally occurring radioisotopes, but there are many synthetic radioisotopes which are manufactured for use in medicine and industry.

Questions

1 The most common isotope of potassium is ^{39}K (93.3% abundance), whereas the most common isotope of uranium is ^{238}U (99.3% abundance). Work out the number of protons and neutrons in each of these isotopes.
2 Which element is commonly used as a fuel in fission reactors?

Alpha (α), beta (β) and gamma (γ) radiation

Many things can happen when a nucleus decays, but in this course, we are concerned with the emission of alpha (α), beta (β) and gamma (γ) radiation.

Alpha

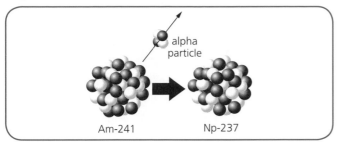

Figure 12.5 Alpha decay of Am-241 to form Np-237 plus an alpha particle

When some radioactive nuclei decay, they eject a 'chunk' of the nucleus consisting of two protons and two neutrons (Figure 12.5). This is known as an **alpha particle** and it is identical to the nucleus of a helium atom. It can be represented as $^4_2$He. The particle carries a positive charge because it has two protons, but no electrons. The charge is not usually included in the symbol of an alpha particle, which can also be written as $^4_2\alpha$. Alpha particles have a mass of 4, but cannot travel very far: they can be stopped by a sheet of paper or a few centimetres of air. We say they have a low penetrating power.

Beta

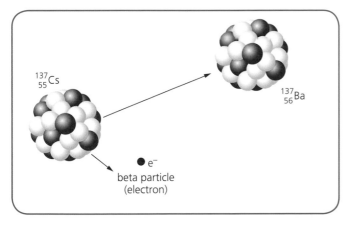

Figure 12.6 Beta decay of Cs-137 to form Ba-137 and an electron

Some nuclei become stable by converting a neutron into a proton and an electron (Figure 12.6):

$$^1_0n \longrightarrow {}^1_1p + {}^0_{-1}e$$

(Note that a proton can also be written as $^1_1$H.)

The proton stays in the nucleus, but the electron is thrown out of the nucleus at high speed and is called a **beta particle** (β). Beta particles are electrons produced *in the nucleus*; they are negatively charged, but they do not come from the orbiting electrons. Beta particles have negligible mass and they have greater penetrating power than alpha particles; they can pass through a sheet of paper, but are absorbed by a thin sheet of metal.

Gamma

Gamma rays ($^0_0\gamma$) are electromagnetic waves which have an even shorter wavelength than X-rays. Gamma rays have no mass and no charge, they are not deflected by electric or magnetic fields, they can travel long distances in air and can only be stopped by thick layers of concrete (Figure 12.7) or lead. Gamma emission does not change the mass or charge of the nucleus; it just releases some excess energy.

Nuclear equations

Nuclear equations are used as a way of showing the changes that take place in the nucleus of radioisotopes; the electrons arranged outside the nucleus are ignored.

In nuclear equations, alpha or beta decay (or emission) means the particle is leaving the isotope, so the symbol for the particle emitted is shown on the right of the arrow. Capture of a particle means that the isotope has absorbed the particle, so the symbol for the particle captured is shown on the left of the arrow.

The following equation shows the alpha decay of americium-241 to form neptunium-237:

$$^{241}_{95}Am \longrightarrow {}^{237}_{93}Np + {}^4_2He$$

In all correct nuclear equations, the sum of the mass numbers (superscripts) on the left of the arrow must equal the sum of the mass numbers on the right of the arrow and the sum of the atomic numbers (subscripts) on the left of the arrow must equal the sum of the atomic numbers on the right of the arrow. In the example given:

$241 = 237 + 4$ (mass numbers)
and $95 = 93 + 2$ (atomic numbers)

Summary

When an isotope emits an alpha particle, the atomic number of the isotope decreases by two and the mass number decreases by four.

When an isotope emits a beta particle, the atomic number of the isotope increases by one (because a neutron has been converted into a proton) and the mass number stays the same.

When an isotope emits gamma rays, the atomic number and the mass number of the isotope stay the same.

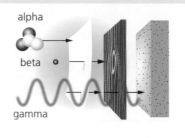

Figure 12.7 The penetrating power of α, β and γ radiation

Questions

3 Copy the diagram below and label the arrows to show which represents α, β and γ radiation. Give a reason for each of your decisions.

Figure 12.8

4 Construct a table showing the name, symbol, mass, charge and penetrating power of α, β and γ radiation.

5 Which of the changes in the table below takes place when an atom emits:
   a) alpha radiation
   b) beta radiation
   c) gamma radiation?

Atomic number	Mass number
A increased	no change
B no change	increased
C decreased	decreased
D no change	no change

Figure 12.9 Marie Curie (left) and Dorothy Hodgkin (right), two celebrated female scientists

Henri Becquerel discovered radioactivity in 1896. Two of Becquerel's colleagues, Marie and Pierre Curie, carried the work forward and one of them, Marie Curie, was the first person to be awarded two Nobel Prizes for different subjects, one for Physics in 1903 and one for Chemistry in 1911. In June 1903 Marie and Pierre were invited to the Royal Institution in London to give a talk on radioactivity. However because Marie was female, she was not allowed to address the meeting and Pierre alone was allowed to speak. Use the internet to research the lives of some famous women scientists of the twentieth and twenty-first centuries.

Worked examples

1 Name the isotope formed when thorium-232 emits an alpha particle.

$$^{232}_{90}Th \longrightarrow {}^{a}_{b}X + {}^{4}_{2}He$$

In this equation: $232 = a + 4$, so $a = 232 - 4 = 228$; $90 = b + 2$, so $b = 90 - 2 = 88$;

If the atomic number is 88, the element is radium, so the isotope is radium-228.

2 Name the isotope formed when lead-212 emits a beta particle.

$$^{212}_{82}Pb \longrightarrow {}^{a}_{b}X + {}^{0}_{-1}e$$

In this equation: $212 = a + 0$ so $a = 212$; $82 = b - 1$, so $b = 83$

If the atomic number is 83, the element X is bismuth, so the isotope is bismuth-212.

3 Name the isotope formed when plutonium-242 captures two alpha particles.

$$^{242}_{94}Pu + 2{}^{4}_{2}He \longrightarrow {}^{a}_{b}X$$

In this equation: $242 + 8 = a$, so $a = 250$; $94 + 4 = b$, so $b = 98$

If the atomic number is 98, the element is californium so the isotope is californium-250.

Questions

6 Find the values of a and b and identify X in each of the following equations:

a) $^{238}_{92}U \longrightarrow {}^{a}_{b}X + {}^{4}_{2}He$

b) $^{32}_{15}P \longrightarrow {}^{a}_{b}X + {}^{0}_{-1}e$

c) $^{220}_{86}Rn \longrightarrow {}^{a}_{b}X + {}^{4}_{2}He$

7 The following sequence of changes occurs as polonium-218 decays.

$$^{218}_{84}Po \xrightarrow{\ 1\ } {}^{214}_{82}Pb \xrightarrow{\ 2\ } {}^{214}_{83}Bi$$

Name the type of radiation emitted during changes 1 and 2. Give a reason for each of your decisions.

8 Copy and complete the following nuclear equations:

a) $^{3}_{1}H \longrightarrow {}^{3}_{2}He$

b) $^{224}_{88}Ra \longrightarrow {}^{228}_{90}Th$

c) $^{60}_{27}Co \longrightarrow {}^{60}_{27}Co$

Half-life

Radioactive decay is a random event. There is no way of knowing when any particular atom will decay; it might happen in the next instant or it might only happen after a million years have passed, so the rate of decay of a radioisotope is measured by a period of time known as the **half-life**.

The time it takes for half the atoms of the starting element to change into new isotopes is called the half-life of that element.

After one half-life has passed, 50% of the atoms have changed into new isotopes. For a particular isotope, the half-life is a constant – that is, it takes the same length of time for 1000 atoms to reduce to 500 as it does for 12 atoms to reduce to 6.

Half-life can also be expressed in terms of activity (as measured on a Geiger counter) or the quantity of the original isotope that is left. The half-life of a particular isotope is fixed and will always be the same whether the isotope is in the Sun (very high temperature) or at the bottom of the ocean (low temperature). Figure 12.10 shows the change in activity for an isotope with a half-life of 2 days.

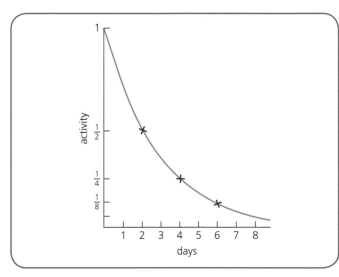

Figure 12.10 Change in activity with time for an isotope of half-life 2 days

Half-lives can vary from seconds to millions of years and are independent of the presence of catalysts or the chemical state of the isotope.

The half-life of the artificial isotope sodium-24 is 15 hours. The particle might be present as an atom,

$^{24}_{11}$Na or as an ion, $^{24}_{11}$Na$^+$, but the half-life would still be 15 hours. A bigger mass of sodium-24 will have a greater *intensity* of radiation than a smaller mass, but the two samples will still have the same half-life. Radioisotopes can be used to date materials because the half-life of an individual isotope is constant.

To examine the random nature of radioactive decay and look at the idea of half-life, place 100 × 2p coins in a plastic beaker, shake the beaker and tip the coins out. Count the number of coins that have landed heads-up, remove them and replace them with 1p coins. (The decayed isotopes have not vanished – they have decayed into a different isotope, represented by the 1p coins.) Repeat the process, counting the number of heads-up 2p coins after each throw and recording the data. When all the 2p coins have been replaced, plot a graph of the data and discuss how this activity relates to the idea of half-life.

Carbon dating

Cosmic radiation bombards the Earth constantly. Neutrons are part of this cosmic radiation and, in the upper atmosphere, nitrogen atoms capture neutrons to produce the radioisotope carbon-14. A proton is also produced in this reaction.

$$^{14}_{7}\text{N} + ^{1}_{0}\text{n} \longrightarrow ^{14}_{6}\text{C} + ^{1}_{1}\text{p}$$

The concentration of carbon-14 atoms in the atmosphere is fairly constant because, as they form, they combine with oxygen and green plants on the planet's surface absorb the CO_2 produced. In living plants, the carbon-14 is slowly but continuously decaying and it is simultaneously being replaced by CO_2 absorbed during photosynthesis. When a plant dies, it stops taking in carbon dioxide, but the carbon-14 already in the plant keeps on decaying. If we assume that the ratio of carbon-14 to carbon-12 in trees today is the same as it was in the past, then measuring the activity due to carbon-14 in a sample of old wood and comparing it to that of a sample of the same type of new wood enables scientists to estimate the age of the old wood.

Working in a group, use the internet to find out why the technique of carbon dating cannot be used to find the age of dinosaur bones. Make a poster to show how carbon dating works.

Worked examples

1 Analysis of a piece of cedar found in Jordan showed the carbon-14 content was 6.25% that of living Lebanese cedar. If the half-life of carbon-14 is 5600 years, estimate the age of the sample.

100% $\longrightarrow$ 50% $\longrightarrow$ 25% $\longrightarrow$ 12.5% $\longrightarrow$ 6.25%
living wood   sample of ancient wood

Each arrow represents one half-life, so four half-lives have passed. The sample is

4 × 5600 = 22 400 years old.

2 Tritium, $^{3}_{1}$H, is a radioisotope of hydrogen with a half-life of 12.4 years. A bottle of old wine showed an activity of 4 counts min^{-1}, whereas a bottle of new wine from the same vineyard had an activity of 32 counts min^{-1}. Calculate the age of the wine.

32 counts $\longrightarrow$ 16 counts $\longrightarrow$ 8 counts $\longrightarrow$ 4 counts
new wine   old wine

Three half-lives have passed, so the wine is
3 × 12.4 = 37.2 years old.

Questions

9 A luminous watch dial containing $^{147}_{61}$Pm (half-life = 2.5 years) has only 1/8th of its original 'glow'. How old is the watch?

10 The activity of a sample of $^{131}_{53}$I was found to be only 1/16th of the activity when it arrived at the hospital 32 days earlier. What is the half-life of iodine-131?

11 Which of the following processes would alter the half-life of a sample of a radioisotope of magnesium?
   A Reacting it with dilute acid
   B Cooling it to $-10\,°C$
   C Burning it in air
   D None of these

Nuclear fission

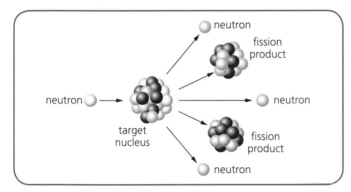

Figure 12.11 Nuclear fission

Nuclear fission (the breaking up of a nucleus) is, at present, the most important way of getting energy from atoms. The main fuel used in nuclear reactors is uranium-235. This is a heavy isotope which can be split into smaller fragments when it is bombarded with neutrons. One pattern of fission is:

$$^{235}_{92}\text{U} + ^{1}_{0}\text{n} \longrightarrow ^{91}_{36}\text{Kr} + ^{142}_{56}\text{Ba} + 3^{1}_{0}\text{n}$$

The fission reaction releases three more neutrons, which collide with more U-235 nuclei and a **chain reaction** takes place. When fission occurs, a tiny quantity of mass is converted into energy according to Einstein's famous equation $E = mc^2$, where c = the speed of light = $3 \times 10^8\,\text{m s}^{-1}$. This equation shows that very small quantities of mass are converted into very large quantities of energy. In fact, the fission of about 235 g of uranium generates the same amount of energy as the burning of 60 tonnes of coal, which would release about 220 tonnes of CO_2 into the atmosphere. There are advantages and disadvantages in using nuclear power to generate electricity, but the discussion is often hampered by the association of the peaceful uses of nuclear energy with nuclear weapons.

Nuclear fusion

Uranium is mined from the Earth's crust and is a finite resource, hence there has been a lot of research into energy production by **nuclear fusion**. Fusion reactions create the energy in stars like our Sun, where very light nuclei fuse (combine) together to make heavier nuclei and release even more energy than fission reactions. For example:

$$^{3}_{1}\text{H} + ^{2}_{1}\text{H} \longrightarrow ^{4}_{2}\text{He} + ^{1}_{0}\text{n} + \text{energy}$$

Fusion has been achieved on this planet, but only for very short time intervals.

Artificial radioisotopes

Radioisotopes (or radionuclides) are manufactured for use in many industries. These can be made by bombarding stable isotopes with neutrons in a nuclear reactor. Neutrons have no charge, so they are not repelled by the positive nucleus.

For example:

$$^{27}_{13}\text{Al} + ^{1}_{0}\text{n} \longrightarrow ^{24}_{11}\text{Na} + ^{4}_{2}\text{He}$$

Uses of radioisotopes

Medicine

1 Cobalt-60 is an artificially produced isotope that carries excess energy. It emits γ-rays to form a lower energy version of the same isotope, and is one of several isotopes used in radiotherapy to destroy cancer cells.
2 Many isotopes are used to trace the movement of materials in our bodies. Salt solution containing a small amount of sodium-24 can be injected into the bloodstream to find out whether blood circulation is normal.
3 Radioisotopes of iodine can be used to test how well the thyroid gland is working, and iodine-131 can be used in radiation therapy for the treatment of thyroid cancer.

Industry

1 Domestic smoke alarms use the α-emitter, americium-241. This isotope is used because the alarms are safe (α-particles have a low penetrating power so are absorbed by the plastic covering) and the isotope has a long half-life (>400 years) so it does not need to be replaced.
2 Beta emitters are widely used to monitor the thickness of materials like paper, plastic and thin metal sheets. Gamma emitters are used to examine castings and welds for imperfections.
3 In many countries food is irradiated, in other words exposed to electron beams, X-rays or gamma rays. The process can be used to kill bacteria that cause food poisoning, such as salmonella and *E. coli*. It can also delay fruit ripening and help stop vegetables such as potatoes and onions from sprouting. Irradiation has little effect on the look and texture of food and though irradiated food has been exposed to radioactivity, it does not become radioactive itself. In the same way that clothes on a washing line exposed to the Sun's rays do not hold the light energy and start to shine, the treated food carries no trace of the high-energy beams it was exposed to.
4 Radioisotopes are used to detect leaks in pipelines and to trace underground water. On big farms, they are often used to check for proper distribution of insecticides.

Checklist for Revision

- I can identify the various radiation processes, alpha, beta and gamma decay.
- I can state the mass, charge and penetrating power of alpha, beta and gamma radiation.
- I can write nuclear equations.
- I can give some uses of radioisotopes.
- I understand the concept of half-life and can use it to solve problems.

End-of-chapter questions

1* Carbon dating uses carbon-14 isotopes and can be used to estimate the age of charcoal found in archaeological sites.
   a) Carbon-14 decays by beta emission. Write a balanced nuclear equation for this decay.
   b) Why can carbon dating *not* be used to estimate the age of fossil fuels?

2 Sulfur-35 is a radioisotope that decays by beta emission.
   a) Write the nuclear equation for the decay of sulfur-35.
   b) The half-life of sulfur-35 is 87 days. How long would it take for the mass of sulfur-35 in an 8 g sample to fall to 2 g?

3 Find the values of a and b and identify X in each of the following equations:
   a) $^{234}_{90}\text{Th} \longrightarrow \,^{a}_{b}\text{X} + \,^{0}_{-1}\text{e}$
   b) $^{210}_{84}\text{Po} \longrightarrow \,^{a}_{b}\text{X} + \,^{4}_{2}\text{He}$
   c) $^{214}_{82}\text{Pb} \longrightarrow \,^{a}_{b}\text{X} + \,^{0}_{-1}\text{e}$

4 Radioisotopes are used in medicine and in industry.
   a) Name a radioisotope used to treat cancer by radiotherapy.
   b) Give two reasons why americium-241 is suitable for use in domestic smoke detectors.
   c) Explain why a γ-emitting radioisotope could not be used to check the thickness of paper as it comes off the rollers in a paper mill.

5 Explain how carbon-14 isotopes are formed and how these isotopes can be used to date samples of wood found in archaeological sites.

6 Everyone on this planet is exposed to background radiation from various sources.
   a) Name two sources of background radiation.
   b) Where were all the naturally occurring elements on this planet formed?

13 Chemical analysis

Consumption and the environment

Figure 13.1 Orville Wright's first successful flight; Neil Armstrong stepping out onto the Moon

In the 66 years between Orville Wright piloting the first successful powered flight in 1903 to Neil Armstrong stepping onto the Moon in 1969 (Figure 13.1), scientists and engineers have transformed and improved our lives beyond belief. Science and scientists have given us cheap electricity and gas to make our lives warm and comfortable sand, more importantly, cheap energy has encouraged the discovery and mass production of

new materials. Factories manufacture (and package) consumer goods, and if the factories are to continue manufacturing, the consumers have to buy more goods. As soon as we started throwing things away, rather than mending them, we created environmental problems and those problems appear, at present, to be increasing and multiplying. Scientists create solutions to problems, but it is up to society whether or not those solutions are accepted. Bio-plastics are being made from corn, rice, potatoes and sugar cane, but these polymers cost a little more and are not always as convenient as traditional oil-based polymers. We consumers like our goods to be cheap and convenient, so our environment continues to be littered with oil-based plastics that can take centuries to decompose.

Water and pollution

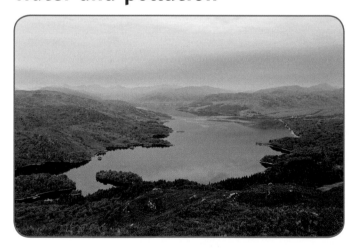

Figure 13.2 Loch Katrine in the district of Stirling

The increasing emphasis on recycling is helping to reduce the mountains of solid material we throw away each year, but water quality is an on-going problem. Scottish Water delivers our clean drinking water (Figure 13.2) and maintains the sewers, drains and water treatment plants. Chemists are employed to monitor and maintain water quality at every step of this efficient process, but industry also needs water.

Distilleries, food processing plants, farms, paper mills, glass factories and just about every other manufacturing industry all use a great deal of water; they then have to clean it before they dispose of it. The Scottish Environmental Protection Agency (SEPA) is the organisation that regulates emissions to our environment. Discharges of water from industrial sites and agriculture are controlled by regulations. Large sites are controlled by licences which limit (among other things) the temperature, pH, conductivity, total suspended solids and the concentrations of nitrate, sodium, potassium, calcium, magnesium, iron, manganese, phosphate and sulfate in this waste water. If a manufacturer breaches this licence, they can be prosecuted through the courts. This means that both SEPA and the manufacturing companies employ chemists to continually check the quality of the water that is being discharged and to 'trouble-shoot' when a problem is found.

Causing and controlling pollution

Figure 13.3 Water pollution can have devastating effects on fish

We often hear people say they don't like 'things with chemicals in them'. This is an odd statement because the substance most essential to life, water, is a chemical and human beings are a mixture of thousands of chemicals, each doing a specific job in our metabolic system. What they mean is that none of us want to find an excess of a particular chemical in a place where it would not occur naturally. For example, how can water become polluted? Here are just two ways:

- Artificial fertilisers can be washed off fields during heavy rain and cause the rapid growth of algae in ponds and lochs. Algae are short-lived. When they die and decay, the decaying process removes oxygen from the water. Fish 'breathe' dissolved oxygen, so the presence of algae can make it difficult for fish to survive.
- A variety of metal compounds can leach into the water from old flooded mine workings.

Each industry has its own type of waste material to add to discharged water. Chemists have developed ways to remove all the unwanted compounds from water, so as long as each industry behaves responsibly, pollution can be avoided.

Techniques for monitoring the environment

Mobile apparatus, such as the Dr Lange photometric analysis kit, can be used to analyse samples taken on site, although it is now usual to send samples to professional laboratories for more detailed and accurate analysis. Although the laboratory analysis is usually automated, the chemical techniques used are often similar to the ones used in school laboratories. Photometric analysis involves adding a reactant to a water sample to produce a coloured solution, passing light of a particular wavelength through this solution and measuring the amount of light absorbed. We can perform some simple experiments in the laboratory in order to help identify unknown substances.

- The presence of some common metals can be shown by carrying out a flame test.
- We can add certain chemicals to a sample of water to identify the presence of both positive and negative ions.
- The process of acid/base titration (see Unit 1) can be used in the laboratory to analyse water samples for acidity or alkalinity.

Flame tests

Some metal compounds will produce a distinctive colour when they are held at the edge of a blue Bunsen flame (Figure 13.4). The colour is due to the presence of the metal ions in the compound. Some of these colours are shown in Table 13.1.

The Royal Society of Chemistry (RSC) has worked with the Nuffield Foundation to produce resources for practical school chemistry. Search the RSC website for a demonstration of the technique of flame testing.

Chemical tests for some positive ions in aqueous solution

Most metal hydroxides are insoluble in water and are formed as precipitates when sodium hydroxide solution is added to a solution containing the metal ion. Several metals can be identified by the colour of their insoluble hydroxides, as shown in Table 13.2.

Figure 13.4 Flame test

Metal ion in solution	Precipitate colour on adding a few drops of sodium hydroxide solution
aluminium	white, but precipitate re-dissolves in excess sodium hydroxide solution
iron(III)	brown
copper	blue
iron(II)	green
magnesium	white

Table 13.2 Identifying metals by the colour of their insoluble hydroxides

(The RSC microscale experiments can be used to minimise the quantities of reactants used for these tests. Details of the microscale procedures can be found in the RSC book *Inspirational Chemistry – Resources for Modern Curricula*, 2006, sections 7.1 to 7.3.)

Chemical tests for some negative ions in aqueous solution

Some common tests for negative ions are shown in Table 13.3.

Element	Ion	Colour
barium	Ba^{2+}	green
calcium	Ca^{2+}	orange-red
copper	Cu^{2+}	blue-green
lithium	Li^+	red
potassium	K^+	lilac
sodium	Na^+	yellow
strontium	Sr^{2+}	red

Table 13.1 Some common flame test results

Ion	Test	Result
chloride, Cl^-	Add a few drops of dilute nitric acid then a few drops of silver nitrate solution.	White precipitate of silver(I) chloride is formed. (Check: this precipitate should be soluble in dilute ammonia solution.)
bromide, Br^-		Cream/off-white precipitate of silver(I) bromide is formed.
iodide, I^-		Yellow precipitate of silver(I) iodide is formed.
carbonate, CO_3^{2-}	Add dilute hydrochloric acid to a solid (or to a solution) containing carbonate ions.	Effervescence (bubbling/fizzing); colourless gas given off – the gas is carbon dioxide, it turns limewater cloudy.
sulfate, SO_4^{2-}	Add a few drops of hydrochloric acid then a few drops of barium chloride solution.	White precipitate of barium sulfate formed – this precipitate is made of very small, fine particles.
nitrate, NO_3^-	Add a few drops of sodium hydroxide solution and a little aluminium powder. Warm the solution in a Bunsen flame and test any gas given off using damp pH paper.	A gas is given off which turns the damp pH paper blue. This shows that the gas is ammonia.

Table 13.3 Some common tests for negative ions

1. Look at the tables of tests (and results) given in this chapter and choose a combination of positive and negative ions to make a compound. On a piece of scrap paper, write down the correct formula for your compound and the results of two tests that will identify the positive and negative ions present in your compound. Ask your teacher/lecturer to check the formula you have written and keep the paper, so you can check the answers. Copy the 'test results' neatly onto a second piece of paper, fold the paper up and put it into a beaker with the test results of the rest of the class. Each member of the class has then got 30 seconds to take a piece of paper from the beaker, identify the compound from the test results and give the correct formula, before the beaker passes on.

2. If time allows, your teacher/lecturer may give you an unknown sample, which might contain a single compound or it might contain a mixture of two compounds. Your task is to plan, then carry out, a series of experiments to identify the ions present in the sample and suggest a name for the compound(s) present.

First, write down an aim for your experiment; this is what you are trying to do (in this case, you are trying to identify an unknown compound). Then write a plan to show the order in which you are going to perform each test and decide how you are going to record the observations. Think about how you are going to ensure your experiments are safe. Ask your teacher/lecturer to check your plan.

When you have finished the practical work, write down what you have found out (this usually goes under the heading 'Conclusion'), then think about how you could have made the process more efficient. If you were going to do the experiment again, would you have done the tests in the same order? Did you make any mistakes? Do you think you correctly identified your compound? If not, why not?

Checklist for Revision

- I know some techniques for monitoring the environment.
- I understand some methods for reducing pollution.
- I can perform calculations based on the results of titrations.

End-of-chapter questions

1. Use the data presented in Table 13.4 to identify compounds A, B, C and D.

Compound	Colour of aqueous solution	Addition of a few drops of acid then of silver(I) nitrate solution	Addition of a few drops of acid then of barium chloride solution	Additional information
A	blue	no reaction	white precipitate formed	blue-green flame test
B	colourless	white precipitate formed	no reaction	lilac flame test
C	colourless	no reaction	white precipitate formed	pH of solution = 1
D	colourless	no reaction	white precipitate which is soluble in dilute ammonia solution	compound D is very soluble in water

Table 13.4

2 A sample of a blue compound, containing copper(II) ions and sulfate ions, was dissolved in water. The solution was divided into three portions and each portion was tested separately. The following results were recorded:

  A Sodium hydroxide solution was added and a blue precipitate was formed.

  B A few drops of hydrochloric acid were added followed by a few drops of barium chloride solution.

  C The compound gave a blue-green colour in a flame test.

  a) Write the formula for the precipitate formed in test A.

  b) Describe what would be seen in test B.

  c) Describe briefly how you would carry out a flame test.

3 Some school students collected a sample of water from their local river and tested it with universal indicator paper; the pH of the water was approximately 10. The students took the sample back to the laboratory and titrated it against a dilute acid because they wanted to find the concentration of hydroxide ions in the water.

  a) Name a piece of laboratory equipment that could be used to give a more accurate pH reading.

  b) i) Name the two pieces of apparatus that are used to measure accurate volumes during a titration.

  ii) Use a pencil and ruler to draw a diagram of the apparatus you would assemble in order to titrate a sample of river water. Label each piece of apparatus and each solution.

  iii)How would you know when the river water had been neutralised?

  c) The results obtained during the titration are shown in Table 13.5. Use the concordant titres to calculate the average titre for this experiment.

	Rough titre	1st titre	2nd titre
Initial burette reading (cm³)	0.0	15.3	29.8
Final burette reading (cm³)	15.3	29.8	44.4
Volume used (cm³)	15.3	14.5	14.6

Table 13.5

  d) This experiment is unlikely to give an accurate value for the concentration of hydroxide ions in the sample. Can you identify a possible source of error in the students' reasoning? Give an explanation for your answer.

Appendix

Practice open-ended questions

1 The Periodic Table groups together elements with similar properties. In most Periodic Tables hydrogen is placed at the top of Group 1 but in some it is placed at the top of Group 7.

   Using your knowledge of chemistry, comment on the reasons for hydrogen being placed either above Group 1 or Group 7.

2 A mass spectrometer is a machine used by scientists to find out information about the isotopes that elements have. Part of the process involves atoms of elements being turned into positive ions.

   Using your knowledge of chemistry, comment on whether the positive ions produced would be bigger, smaller or the same size as the atoms from which they were made.

3 Many different hydrocarbon compounds exist. A student was given three bottles labelled A, B and C and told that one bottle contained hexane, another bottle contained cyclohexane and the third bottle contained hex-1-ene.

   Using your knowledge of chemistry, describe how you would correctly identify the contents of each bottle.

4 Esters, alcohols and carboxylic acids are widely used in everyday life. A student was given three bottles labelled A. B and C and told that one bottle contained an ester, another bottle contained an alcohol and the third bottle contained a carboxylic acid.

   Using your knowledge of chemistry, describe how you would correctly identify the contents of each bottle.

5 'Mag Wheels' were a popular type of alloy wheel fitted to sports cars in the 1950s and 1960s. The wheels were made using an alloy of magnesium, and they were preferred to steel wheels by manufacturers of racing cars and road-going sports cars. Today, most of these cars have aluminium alloy wheels.

   Using your knowledge of chemistry, comment on why manufacturers may have used 'Mag Wheels' instead of steel ones on sports cars, and why today an aluminium alloy might be used instead.

Practice assignments

1 Splitting the atom

The name atom comes from the Greek work 'atomos' which means indivisible or something that cannot be divided further.

During the late nineteenth century and early twentieth century, scientists discovered particles inside the atom thereby demonstrating that the atoms were, in fact, divisible and that the name 'atom' may not be appropriate.

These discoveries resulted in scientists looking at 'splitting the atom' and using the energy produced in this highly exothermic reaction.

Using research you have done, produce a report that outlines briefly what is meant by splitting atoms and identifies which scientists are credited with the first 'splitting' of the atom. You should also provide a balanced evaluation of the pros and cons of using nuclear power.

Make sure you include details of some of the chemistry involved and, at the end of the piece, list all the websites and other sources you have looked at during your research, even if you have not quoted from them directly.

2 Acid rain

Acid rain has been an identified problem in Norway since the late 1970s, and while awareness of the causes of acid rain has increased and steps have been taken to reduce the spread of the problem, there is still evidence of pollution attributed to acid rain in Scandinavian countries.

Imagine you are a Scottish reporter investigating the main causes of acid deposition in Scandinavia. Use the internet to research the damage that is being caused by acid rain. What are the sources of these problems? Which country/countries is/are responsible for the origin of the pollutant gases damaging the environment? Who do you think should be responsible for the burden of cleaning up the environment in Norway?

Using your knowledge of chemistry, suggest how the effects of acid rain can be further reduced, providing alternatives to any harmful substances or processes involved. What would be the financial implications of making your suggested changes? Try to draw a reasonable conclusion about how the world has to change its practices to stop the environmental pollution caused by acid rain.

Make sure you include details of some of the chemistry involved and, at the end of the piece, list all the websites and other sources you have looked at during your research, even if you have not quoted from them directly.

3 Alcohol

People have been producing and consuming alcohol for thousands of years. The alcohol that people drink (ethanol) is made by fermenting fruit and cereals, so it is not particularly expensive to produce, and it can be distilled easily to make the alcohol solution more concentrated. Alcohol-related illness costs the NHS in Scotland over £100 million each year, but the Government makes billions of pounds each year by taxing alcohol. Should we, in Scotland, continue to allow cheap alcohol to be sold in every supermarket?

Write a short balanced piece for an internet blog about the impact of alcohol on society; include a brief explanation of distillation, put the arguments for and against the supply of cheap and easily available alcohol and draw a reasoned conclusion.

List all the websites and other sources you have looked at during your research, even if you have not quoted from them directly.

4 Methanol as a fuel

Methanol is used in some countries as a fuel for racing cars mainly because it is less volatile and explosive than petrol. Methanol is also cheaper to produce and when it burns, it undergoes almost complete combustion, so it produces less carbon monoxide than petrol.

Use the internet to find out why methanol is not more widely used as a fuel and write a short article for a popular newspaper giving the arguments for and against using methanol as a fuel for cars. Include some chemistry (for example, explain what complete combustion means) and draw a reasoned conclusion.

Remember to list all the websites and other sources you have looked at during your research, even if you have not quoted from them directly.

5 Biomass

Biomass is the name given to any organic material that can be used to obtain energy. Sweden gets over 30% of its energy from biomass technology and it is also widely used in Norway and Denmark. The Highlands and Islands are investing in it (a biomass plant has opened on the Cromarty Firth) and some distilleries are looking at ways of using their waste material to help them become self-sufficient energy-wise. Domestic waste can be used as biomass, so why build massive wind farms, which only produce electricity on a windy day and which are expensive to maintain, when we can make energy from potato peelings?

Use the internet to research arguments for and against wind power and biomass technology. Write a short article giving the arguments for both forms of renewable energy and draw a reasoned conclusion.

Remember to list all the websites and other sources you have looked at during your research, even if you have not quoted from them directly.

6 SMAs

Alloys are mixtures of a metal and another element, usually another metal. Shape memory alloys (SMAs) can 'remember' their original shape. When they are heated above a certain temperature (the transition temperature), they will return to that shape no matter how much they have been bent or folded. They do this because at their transition temperature, SMAs change from one type of crystal structure to another, and when they change their structure, they change their shape.

Using the internet for research, write a short, balanced article for a scientific magazine on two examples of how and where SMAs are used and their impact on society and the environment.

Make sure you include brief details of the chemistry involved and at the end of the piece, list all the websites and other sources you have looked at during your research, even if you have not quoted from them directly.

7 Rechargeable batteries

Silver–cadmium rechargeable batteries are small but very expensive; they are often found in satellites where space and weight are important.

Using the internet for research, write a short, balanced article for an online magazine on two examples of how and where rechargeable batteries are used and their impact on society and the environment.

Make sure you include details of some of the chemistry involved and, at the end of the piece, list all the websites and other sources you have looked at during your research, even if you have not quoted from them directly.

8 Fuel cells

For the past few years, the forklift trucks used in all Walmart and Pepsi™ warehouses have been powered by fuel cells.

Use the internet to research and write a short, balanced article for a scientific magazine on the applications of modern hydrogen fuel cells and their impact on society and the environment.

Make sure you include details of some of the chemistry involved (including at least one method of obtaining the hydrogen) and at the end of the piece, list all the websites and other sources you have looked at during your research, even if you have not quoted from them directly.

9 Food insecurity and artificial fertilisers

Use the website of the Food and Agricultural Organisation of the United Nations (given below) to find out about present levels of food insecurity in the world, and search the internet to find out what a man called Thomas Malthus wrote in 1798, what the 'Green Revolution' means and what 'Nitrate Vulnerable Zones' (NVZs) are.

www.fao.org

Write a short, balanced article for a scientific magazine on the applications of artificial fertilisers and their impact on society and the environment.

Make sure you include details of some of the chemistry involved and, at the end of the piece, list all the websites and other sources you have looked at during your research, even if you have not quoted from them directly.

10 Fritz Haber and the First World War

The History Department in your school may be able to help you with this task. In 1900, Europe was dependent on supplies of nitrates shipped from South America. Nitrates were essential for agriculture and for the manufacture of munitions and explosives, so the German corporation, BASF, started to look for a way to manufacture nitric acid. In 1908, Haber entered into an agreement with BASF to work with Karl Bosch towards the industrial-scale production of ammonia as a feedstock for the Ostwald process.

Use the internet to research the possible connection between the large-scale production of ammonia and the outbreak of the First World War. Include some details of the chemistry; this could involve the manufacture of ammonia or nitric acid, fertilisers or explosives etc. Give a balanced evaluation (in other words, give the arguments on both sides) of the rather extreme statement, 'the Haber process was responsible for the First World War' and draw a reasoned conclusion from your research.

Remember to list all the websites and other sources you have looked at during your research, even if you have not quoted from them directly.

11 GM foods

The Soil Association (founded in 1946) campaigns on behalf of organic farmers and is against the production of GM (genetically modified) foods. Organic farmers do not use artificial fertilisers or pesticides; they only use organic (natural) fertilisers to replace lost nutrients.

Organic foods cost more and have to be washed carefully as they can carry bacteria such as *E. coli.* Many scientists claim that GM crops and artificial fertilisers are needed if the poorest people on our planet are to survive.

Research both sides of this argument and set up a class discussion with two or three volunteer 'experts' speaking for each side and a chairperson to manage the debate. Invite questions from the class, then have a vote to see which side has 'won'.

Make sure you include details of some of the chemistry involved and list all the websites and other sources you have looked at during your research, even if you have not quoted from them directly.

12 Radioisotopes

Radioisotopes are made in nuclear reactors and are widely used in medicine and industry.

Use the internet, and this textbook, to investigate how two different radioisotopes are made and what they are used for.

Many people are afraid of nuclear power. Write a short, reasoned article on the use of radioisotopes balanced against the way in which they are made.

Draw a conclusion at the end of your article, and remember to list all the websites and other sources you have looked at during your research, even if you have not quoted from them directly.

13 Fluoride in drinking water

In 2004, the first minister of Scotland 'pulled the plug' on plans to add fluoride to drinking water.

A lot of districts in England have very small quantities of fluoride added to the drinking water in an attempt to reduce tooth decay in children. There has been much debate on the subject, but no fluoride is added to drinking water in either Scotland or Wales. Fluoride occurs naturally in tea, fish, seawater and (in some regions) drinking water.

Use the internet to research this topic and write a short article for a website about the deliberate addition of fluoride to drinking water. Give a balanced evaluation of the impact on society, particularly on children, of adding fluoride and draw a reasoned conclusion, for or against.

Make sure you include details of some of the chemistry involved and, at the end of the piece, list all the websites and other sources you have looked at during your research, even if you have not quoted from them directly.

14 Chemical analysis

You are an environmental chemist and you have been asked to analyse a sample of soil from a brownfield site. You know that the site used to be a scrap metal yard, so possible contaminants could be iron(III) ions (from rusty metal) and sulfuric acid from old car batteries.

Use the internet and your own knowledge to write a short scientific report on the tests you intend to do and assume that your tests subsequently prove to be positive. Discuss the problem of soil contamination and the impact this has on the environment and on building costs. Has the increase in sensitivity of analytical tools had an effect on legislation concerning acceptable levels of contamination? What are the costs of clearing contaminated soil from building sites? What can you conclude about the health of the population now, when pollution is a political issue, compared to 80 years ago, when it was not?

Make sure you include details of some of the chemistry involved, and at the end of the piece, list all the websites and other sources you have looked at during your research, even if you have not quoted from them directly.

Glossary

A

Acid: A substance that produces hydrogen ions (H^+) when dissolved in water.

Addition polymerisation: The reaction between many small, unsaturated molecules (monomers) to form one large molecule (a polymer) and nothing else.

Addition reaction: A chemical reaction where molecules add across a double bond.

Alcohol: A compound containing an –OH group.

Alkali: A soluble base that produces hydroxide ions (OH^-) when dissolved in water.

Alkanes: A family of hydrocarbons where all the members are saturated and have the general formula C_nH_{2n+2}.

Alkenes: A family of hydrocarbons where all the members are unsaturated, they contain one carbon-to-carbon double bond and have the general formula C_nH_{2n}.

Alpha (α) particle: A charged particle consisting of two protons and two neutrons emitted by some radioisotopes. It is identical to a helium nucleus and is represented as ^4_2He.

Atom: The smallest quantity of an element that can take part in a chemical reaction.

Atomic lattice: A large three-dimensional arrangement of atoms held together by covalent bonds.

Atomic number: The number of protons in the nucleus of an atom of an element.

B

Base: A substance that will neutralise an acid.

Battery: A series of chemical cells joined together (often the words cell and battery are used interchangeably).

Beta (β) particle: Charged particle consisting of a single electron emitted from the nucleus of some radioisotopes. Represented as $^0_{-1}e$.

Biodegradable: Able to rot away by natural biological processes.

Burette: A graduated piece of apparatus used to measure accurately the volume of liquid or solution dispensed during an experiment.

C

Calorimeter: Equipment used to measure the energy released in a chemical reaction.

Carboxylic acid: An acidic compound that contains the carboxyl group (–COOH).

Catalyst: A substance that will speed up a chemical reaction but is not used up during the process.

Cell: In a chemical cell chemical energy is converted into electrical energy.

Ceramic: A compound of a metal and a non-metal that has gained certain properties by being heated and then cooled.

Compound: A substance made up of two or more elements chemically joined.

Concentration: A measure of how much solute is dissolved in a solvent, measured in $mol\,l^{-1}$.

Condensation polymerisation: A reaction in which many small monomer molecules join to form one large polymer molecule with water, or some other small molecule, formed at the same time.

Condensation reaction: One in which two or more molecules join to produce a single larger molecule with water, or another small molecule, produced at the same time.

Covalent bond: A shared pair of electrons between two non-metal atoms.

Covalent network: A giant network of non-metal atoms held together by covalent bonds.

Cycloalkanes: A family of hydrocarbons where all the members are saturated and have the general formula C_nH_{2n}.

D

Dilution: Addition of more solvent to a solution to decrease its concentration.

Dissociation: A process in which compounds separate into oppositely charged ions.

E

Electrochemical series (ECS): A list of metals (and hydrogen) in order of their ability to lose electrons and form ion in solution.

Electrolyte: A compound that conducts owing to the movement of ions either when dissolved in water or melted.

Electron: A negatively charged particle with a relative mass of zero, which orbits the nucleus of an atom.

Element: A substance that consists of atoms with the same number of protons in their nuclei.

Endothermic reaction: A reaction in which energy is taken in from the surroundings.

Enzyme: A biological catalyst.

Ester: A fruity smelling compound used in fragrances, flavourings and as a solvent. It is formed by reacting an alcohol with a carboxylic acid.

Exothermic reaction: A reaction in which energy is released to the surroundings.

F

Feedstock: A substance obtained from a raw material which is then used to manufacture another substance.

G

Gamma (γ) radiation: High-frequency, and high-energy, electromagnetic radiation emitted by radioactive substances.

Group ion: A charged particle that contains more than one type of atom (such as permanganate, MnO_4^-).

H

Half-life: The time in which the activity of a radioisotope decays by half, or in which half of its atoms disintegrate.

Homologous series: A family of compounds with the same general formula and similar chemical properties that show a gradual change in physical properties. The alkanes are an example of a homologous series.

Hydration: An addition reaction where water is added across a double bond.

Hydrogenation: An addition reaction where hydrogen is added across a double bond.

I

Indicator: A substance whose colour changes depending on pH.

Ion: A charged particle.

Ion–electron equation: Equation that shows either the loss of electrons (oxidation) or the gain of electrons (reduction).

Ionic bond: The electrostatic force of attraction between positive ions and negative ions.

Ionic lattice: A large arrangement of ions held together by ionic bonds.

Isomers: Compounds with the same molecular formula but different structural formulae.

Isotopes: Atoms with the same atomic number but different mass numbers.

M

Macroscopic: Large enough to be seen or examined by the unaided eye.

Malleable: A physical property of metals. A malleable material can be shaped by hammering or rolling.

Mass number: Equal to the number of protons plus neutrons in an atom or ion.

Metallic bonding: Each atom in the metal element loses its outer electrons to form positive ions. These ions pack together in a regular crystalline arrangement and the free (delocalised) electrons move freely through the structure. The electrostatic attraction of the positive ions for the delocalised electrons binds the ions together and is called metallic bonding.

Mole: The relative atomic mass of an element expressed in grams (twice the RAM for diatomic elements); the relative formula mass of a compound expressed in grams.

Molecule: Two or more atoms held together by covalent bonds.

Molten: Describes a liquid formed by melting a solid.

Monomers: Relatively small molecules that can join together to produce a very large molecule (a polymer) by a process called polymerisation.

N

Neutralisation: A reaction in which the pH moves towards 7.

Neutron: A neutral particle with a relative mass of 1 found in the nucleus of an atom.

Nuclear fission: Splitting nuclei by bombarding them with slow-moving neutrons.

Nuclear fusion: Light nuclei combining to form heavier nuclei and producing a lot of energy.

Nucleus: The positively charged centre of an atom which contains the neutrons and protons.

O

Oxidation reaction: One in which electrons are lost or oxygen is gained.

P

pH: A measure of how acidic or alkaline a substance is.

Phenyl group: A group of carbon and hydrogen atoms, formula C_6H_5-.

Pickling: A method used to keep food from spoiling. It involves storing the food in vinegar, which contains ethanoic acid, to stop bacteria and fungi growing.

Piezoelectric: Crystals that acquire an electrical charge when twisted, distorted, compressed or put under any mechanical stress are said to be piezoelectric.

Pipette: A piece of apparatus used to measure a specific volume of liquid or solution.

Polymer: A very large molecule formed by the joining together of many smaller molecules (monomers).

Proton: A positively charged particle with a relative mass of 1 found in the nucleus of an atom.

R

Radioisotopes: Atoms that emit radiation because their nuclei are unstable. Three types of radiation, α, β or γ, may be emitted.

Rate: A measure of how quickly a chemical reaction is progressing.

Reducing agent: A substance that gives electrons away to another substance or that removes oxygen from it.

Reduction: A process in which electrons are gained or oxygen is lost.

Relative atomic mass (RAM): The average mass of the isotopes of an element.

Relative formula mass (RFM): The sum of the relative atomic masses of all the elements in a compound.

S

Saturated: A compound that cannot have any more atoms added to it. All the bonds are single.

Spectator ion: An ion that is present as both a reactant and a product in a chemical equation. Spectator ions take no part in the reaction.

Systematic name: The chemical name for a compound, obtained by following a set of prescribed rules.

T

Tetrahedral: A tetrahedral molecule has a central atom bonded to four other atoms that are positioned at the corners of a regular tetrahedron.

Thermoplastic: A plastic that softens on heating and can be reshaped.

Titration: An experiment that will accurately measure the volume of liquid or solution required for a particular reaction.

U

Unsaturated: A compound that can have more atoms added to it. It will contain at least one carbon-to-carbon double bond.

V

Valency: The number of bonds that an atom or ion can form.

Variable: A factor that can be changed in a chemical reaction that will affect the rate.

Index

Index

Answers

Chapter 1 Rate of reaction

End-of-chapter questions

1 A

2 A

3 B

4 A

5 B

6 A powdered catalyst will have a greater surface area. This will allow for a greater number of successful collisions to occur on the surface of the catalyst molecule.

7 a)

copper(II) carbonate and dilute hydrochloric acid

balance used to measure the loss

127.46 g

b)

c) The line that is plotted becomes horizontal, indicating that even though time is progressing, there is no increase in the mass of gas that is produced.

d) $0.024\,g\,s^{-1}$

8 a) Hydrogen gas will burn with a squeaky pop.

b) Any two from the following three: use a lower concentration of hydrochloric acid; use larger particles of zinc; lower the temperature of the reaction mixture.

c)

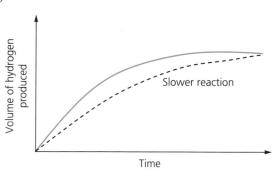

Slower reaction

9 a) $0.0175/0.018\,mol\,l^{-1}\,min^{-1}$

b) $0.004\,mol\,l^{-1}\,min^{-1}$

10 a)

b) $55\,cm^3$

c)

d)

carbon dioxide collected

chalk lumps + HCl

Chapter 2 Atomic theory

End-of-chapter questions

1 A
2 B
3 A
4 B
5 D
6 a)

Particles	Number
protons	47
electrons	46
neutrons	60

b) Isotopes
c) Equal quantities of each isotope.

7 a)

Isotope	Number of protons	Number of electrons	Number of neutrons
$^{35}_{17}Cl$	17	17	18
$^{37}_{17}Cl$	17	17	20

b) The isotope with a mass of 35 is the more abundant.

8

Isotope	Mass number	Atomic number	Number of protons	Number of electrons	Number of neutrons
$^{31}_{15}P$	31	15	15	15	16
$^{39}_{19}K$	39	19	19	19	20
$^{15}_{7}N$	15	7	7	7	8

9 a) $^{41}_{20}Ca$
b) Protons = 20; neutrons = 21
c) Equal numbers of positive protons and negative electrons

10

Ion	Number of protons	Number of electrons	Number of neutrons
$^{35}_{17}Cl^-$	17	18	18
$^{16}_{8}O^{2-}$	8	10	8
$^{40}_{20}Ca^{2+}$	20	18	20

Chapter 3 Bonding, structure and properties

End-of-chapter questions

1 D
2 B
3 B
4 C
5 a) A covalent bond is a shared pair of electrons between two non-metals. The atoms are held together because of the electrostatic force of attraction between the positive nuclei of each atom and the negatively charged electrons.

b)

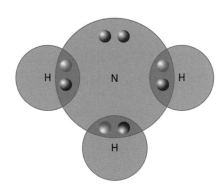

c)

6

Substance	Bonding and structure
C	ionic
B	covalent network
A	covalent molecular

7 Silicon dioxide has a covalent network structure. This means that a lot of energy is required to break the strong covalent bonds that hold the atoms together, meaning that very high temperatures are required to turn silicon dioxide into a gas. Carbon dioxide has a covalent molecular structure. This means that only the weak forces of attraction between the molecules are broken when it is turned into a gas. This is why it is a gas at room temperature.

8 Salt is an ionic compound. In solution, the ions are free to move. Glucose is a covalent molecular compound. Since molecules are not charged particles, a current cannot flow.

Chapter 4 Chemical formulae

In-text questions

1 a) CO
   b) BBr_3
   c) PCl_5
   d) N_2O_3
   e) XeF_6
2 a) Li_2S
   b) KCl
   c) Ca_3P_2
   d) BBr_3
   e) Mg_3N_2
3 a) $Mg(OH)_2$
   b) $NaHCO_3$
   c) $KMnO_4$
   d) $Al_2(SO_4)_3$
   e) $(NH_4)_3PO_4$
4 a) $Mg^{2+}(OH^-)_2$
   b) $Na^+HCO_3^-$
   c) $K^+MnO_4^-$
   d) $(Al^{3+})_2(SO_4^{2-})_3$
   e) $(NH_4^+)_3PO_4^{3-}$
5 a) Au_2O_3
   b) $SnCl_4$

   c) Fe_2S_3
   d) $CuBr$
   e) PbI_2
6 a) $C_2H_4 + 3O_2 \longrightarrow 2CO_2 + 2H_2O$
   b) $3Ag_2S + 2Al \longrightarrow Al_2S_3 + 6Ag$
   c) $C_6H_{12}O_6 + 6O_2 \longrightarrow 6CO_2 + 6H_2O$
   d) $2H_3PO_4 + 3Mg(OH)_2 \longrightarrow Mg_3(PO_4)_2 + 6H_2O$
   e) $2CO + O_2 \longrightarrow 2CO_2$
7 a) $64\,g$
   b) $16\,g$
   c) $158\,g$
   d) $100\,g$
   e) $148.5\,g$
8 a) $186\,g$
   b) $84\,g$
   c) $11.5\,g$
   d) $245\,g$
   e) $477.5\,g$
9 a) 3 moles
   b) 1.5 moles
   c) 0.1 moles
   d) 2.5 moles
   e) 4 moles
10 a) $0.25\,g$
   b) $1.4\,g$
   c) $2.2\,g$
   d) $90\,g$
11 a) $0.5\,mol\,l^{-1}$
   b) $1.5\,mol\,l^{-1}$
   c) $1\,mol\,l^{-1}$
   d) $2\,mol\,l^{-1}$
   e) $2.4\,mol\,l^{-1}$
12 a) $0.2\,mol\,l^{-1}$
   b) $1\,mol\,l^{-1}$
   c) $0.4\,mol\,l^{-1}$
   d) $2\,mol\,l^{-1}$
   e) $0.5\,mol\,l^{-1}$

Chapter 5 Acids and bases

End-of-chapter questions

1 C
2 D
3 C
4 C
5 a) Pipette
   b) Neutralisation

c) 0.0025 moles

d) 24.8 cm³

6 a) potassium carbonate + sulfuric acid $\rightarrow$
  potassium sulfate + water + carbon dioxide

b) $K_2CO_3 + H_2SO_4 \rightarrow K_2SO_4 + H_2O + CO_2$

c) 2 mol l⁻¹

7 a) A base is a substance that will neutralise an acid.

b) An alkali is a soluble base.

8 0.12 mol l⁻¹

9 0.025/0.03 mol l⁻¹

10 0.143/0.14 mol l⁻¹

11 60 cm³

12 16 cm³

13 0.25 mol l⁻¹

14 0.0417/0.04 mol l⁻¹

15 300 cm³

16 300 cm³

17 3

Chapter 6 Homologous series

In-text questions

1 a)

b)

c)

End-of-chapter questions

1 a) B and C

b) B and E

c) E

d) A, D and F

e) D

2 a) A and E

b) A

c) F

d) C

e) D and F

3 a)

b)

c)

4 A

B

5 a)

b)

6 a) 2-methylhexane

b) 4,4-dimethylpent-2-ene

c) 2,3-dimethylbut-1-ene

7 a) C_nH_{2n+2}

b) i) $C_{12}H_{26}$

ii) $C_{39}H_{80}$

8 a)

Name	Structure	Molecular formula
ethyne	$H-C \equiv C-H$	C_2H_2
propyne	$H-C \equiv C-\overset{\displaystyle H}{\underset{\displaystyle H}{C}}-H$	C_3H_4
but-1-yne	$H-C \equiv C-\overset{\displaystyle H}{\underset{\displaystyle H}{C}}-\overset{\displaystyle H}{\underset{\displaystyle H}{C}}-H$	C_4H_6

b)

c) C_nH_{2n-2}

9 a) i) $C_nH_{2n}O$

ii) If the OH is on the end/first carbon atom, an aldehyde is obtained. If the OH is on the second carbon atom, a ketone is obtained.

b)

c)

10 a) A family of compounds with the same general formula and similar chemical properties that show a gradual change in physical properties.

b) C_nH_{2n-2}

c) $C_5H_8Br_4$

d)

Chapter 7 Everyday consumer products

In-text questions

1 a)

b)

2 Hexan-5-ol would be incorrect because alcohols are named so that the –OH is attached to the lowest numbered carbon atom. Hexan-5-ol should be called hexan-2-ol.

3 a) Pentan-2-ol

b) Hexan-3-ol

4 a)

b)

5 a) Pentanoic acid

b) Heptanoic acid

6 $C_nH_{2n}O_2$

End-of-chapter questions

1 C

2 a) D

b) C and D

c) D

3 B An unsaturated carboxylic acid

4

5

6 Can be used to make esters. Can be used as cleaning chemicals. Can be used as food preservatives.

7 $C_2H_5OH + 3O_2 \longrightarrow 2CO_2 + 3H_2O$

8 Fragrances, flavourings and solvents

9 a)

   b)

10 a) Add bromine water to a sample of the lavender oil and shake. If it contains unsaturated compounds, the bromine water will decolourise.

   b) –OH

   c) It is an ester.

   d)

Chapter 8 Energy from fuels

In-text questions

1 a) $CH_4 + 2O_2 \longrightarrow CO_2 + 2H_2O$

   b) $C_4H_{10} + 6\frac{1}{2}O_2 \longrightarrow 4CO_2 + 5H_2O$

   c) $C_6H_{14} + 9\frac{1}{2}O_2 \longrightarrow 6CO_2 + 7H_2O$

2 a) $C_2H_5OH + 3O_2 \longrightarrow 2CO_2 + 3H_2O$

   b) $C_3H_7OH + 4\frac{1}{2}O_2 \longrightarrow 3CO_2 + 4H_2O$

   c) $C_6H_{13}OH + 9O_2 \longrightarrow 6CO_2 + 7H_2O$

3 72 g

4 176 g

5 4.5 g

6 44 g

7 76.67/76.7 g

8 64 g

9 14.28/14.3 g

10 10 g

End-of-chapter questions

1 a) Exothermic; 60 kJ released

   b) Endothermic; 70 kJ taken in

2 180 g

3 1760 g

4 98.29/98.3 g

5 13.75 g

6 47.9 g

7 297.2 kJ

8 1933.25/1933.3 kJ

9 a) Measure the temperature of the water at the start of the experiment and measure the highest temperature of the water. Measure the mass of the spirit burner before burning and after burning. Measure the volume/mass of water.

   b) 208.65/208.6 kJ

   c) Less heat is lost to the surroundings and the combustion is more complete.

10 145.45 kg

Chapter 9 Metals

In-text questions

1 a) The electrostatic attraction of positive metal ions for delocalised electrons present in the lattice.

   b) Malleable means something that can be shaped by hammering or rolling.

2 Agree. Covalent bonding involves sharing of electrons between bonded atoms and all the metal ions in the lattice share their outer (delocalised) electrons. Ionic bonding involves loss and gain of electrons and metal atoms in the lattice lose their outer electrons, which become delocalised.

3 a) Mercury, Hg

   b) Lattice means a regular three-dimensional arrangement of particles.

   c) The metal ions in the lattice change places and slide past each other to adapt to a new shape. This happens because all the ions in a metal lattice are the same size and have the same charge.

4 a) $4K + O_2 \longrightarrow 2K_2O$
   b) $4K + O_2 \longrightarrow 2(K^+)_2O^{2-}$
5 a) $Ca + 2H_2O \longrightarrow Ca(OH)_2 + H_2$
   b) $Ca(s) + 2H_2O(l) \longrightarrow$
      $$Ca^{2+}(aq) + 2OH^-(aq) + H_2(g)$$
6 a) $Mg + 2HCl \longrightarrow MgCl_2 + H_2$
   b) $Mg(s) + 2H^+(aq) + 2Cl^-(aq) \longrightarrow$
      $$Mg^{2+}(aq) + 2Cl^-(aq) + H_2(g)$$
7 a) C A B
   b) C = K, Na, Li, Ca or Mg; A = Al, Zn, Fe, Sn or Pb; B = Cu, Hg, Ag or Au
8 a) Group 3
   b) GaN
   c) Silicon
9 Any two elements, with correct symbols, from either the Lanthanide or Actinide series.
10 a) $2HgO \longrightarrow 2Hg + O_2$
   b) $2Hg^{2+}O^{2-}(s) \longrightarrow 2Hg(l) + O_2(g)$
11 a) $2CuO + C \longrightarrow 2Cu + CO_2$
   b) $2Cu^{2+}O^{2-}(s) + C(s) \longrightarrow 2Cu(s) + CO_2(g)$
   c) Carbon
12 72.41/ 72.4%
13 52.94/52.9%
14 a) 39.31/39.3%
   b) 56.40/56.4%
   c) 52.19/52.2%
   d) 38.71/38.7%
15 a) Oxidation
   b) Reduction
   c) Reduction
16 B
17 a) $Fe(s) \longrightarrow Fe^{2+}(aq) + 2e^-$ [O]
      $2H_2O(l) + O_2(g) + 4e^- \longrightarrow 4OH^-(aq)$ [R]
   b) $2Fe(s) + 2H_2O(l) + O_2(g) \longrightarrow$
      $$2Fe^{2+}(aq) + 4OH^-(aq)$$
18 a) From Mg to Fe through the wires and meter
   b) From Sn to Ag
   c) From Zn to Ni
   d) From Al to Pb
19 a) Fe/Cu
   b) Fe/Ag
   c) Mg/Au
   d) Al/Hg
20 a) The electrolyte completes the circuit.
   b) The reactants are used up.
21 6

End-of-chapter questions

1 a) $2PbO + C \longrightarrow 2Pb + CO_2$
   b) Magnesium is too reactive.
   c) Mercury, silver or gold
2 a) The air/atmosphere or electrolysis of water
   b) Fuel cells cause less pollution than petrol/petrol comes from a finite resource (crude oil) but hydrogen can be obtained by electrolysis of water using energy from renewable sources.
   c) $H_2(g) \longrightarrow 2H^+(aq) + 2e^-$
3 a) Electrons flow from X to Y.
   b) Reduction
   c) $2Fe^{3+}(aq) + SO_3^{2-}(aq) + H_2O(l) \longrightarrow$
      $$2Fe^{2+}(aq) + SO_4^{2-}(aq) + 2H^+(aq)$$
4 a) 3+
   b) $Al^{3+} + 3e^- \longrightarrow Al$
5 B
6 a) Electrode A/positive electrode
   b) $2Na^+ + 2H^- \longrightarrow 2Na + H_2$
   c) Metal elements contain delocalised electrons that carry the current.
7 A
8 a) $Zn(s) + Cu^{2+}(aq) \longrightarrow Zn^{2+}(aq) + Cu(s)$
   b) i) $Fe(s) \longrightarrow Fe^{2+}(aq) + 2e^-$
      ii) Oxidation

Chapter 10 Properties of plastics

In-text questions

1 a) Addition polymer; the polymer is a hydrocarbon so the backbone chain must only contain carbon atoms or is an alkene, or is unsaturated.
   b) Buta-1,3-diene
2 a) Condensation polymer; the backbone chain contains oxygen atoms.
   b) Repeating unit is ($-O-CH_2-CH_2-$); full structural formula is

3 a)

b)

4 a) Propene and poly(propene)

b) Alkenes

5

6 a) An addition polymer is a very big molecule formed when many small, unsaturated molecules (monomers) combine to form the polymer and nothing else.

b)

End-of-chapter questions

1 a)

b)

c) i) Biodegradable means a polymer is able to rot away by natural biological processes.

ii) To speed up the decomposition of plastics in landfill sites and reduce pollution.

2 a) Condensation polymerisation

b)

3 a)

b) i) An ester

ii)

4 a) An addition polymer

b)

5 a) A smart material is one that changes shape in response to changes in its environment.

b) Hair gel/babies' nappies/water crystals for plants/ Jell-O snacks/dressings for wounds/contact lenses

c) Silk/wool/hair/fur/rubber/DNA/starch/cellulose

Chapter 11 Fertilisers

In-text questions

1 $Ca(OH)_2 + (NH_4)_2SO_4 \longrightarrow CaSO_4 + 2NH_3 + 2H_2O$

2 Dimerise means two monomer units combining together.

3 N_2O_4

4 Dinitrogen monoxide

5 $NO_2 + O_2 \longrightarrow NO + O_3$; O_3 is called ozone; it forms in the upper atmosphere and absorbs some of the harmful ultra-violet radiation emitted by the Sun.

6 To increase the surface area and make the catalyst more efficient.

7 $4NH_3(g) + 5O_2(g) \longrightarrow 4NO(g) + 6H_2O(g)$
$2NO(g) + O_2(g) \longrightarrow 2NO_2(g)$
$4NO_2(g) + O_2(g) + 2H_2O(l) \longrightarrow 4HNO_3(aq)$

8

Acid	Formula of acid	Ions present in aqueous solution
phosphoric	H_3PO_4	$3H^+(aq)$ and $PO_4^{3-}(aq)$
sulfuric	H_2SO_4	$2H^+(aq)$ and $SO_4^{2-}(aq)$
nitric	HNO_3	$H^+(aq)$ and $NO_3^-(aq)$
hydrochloric	HCl	$H^+(aq)$ and $Cl^-(aq)$

9 $Ca_3(PO_4)_2 + 3H_2SO_4 \longrightarrow 2H_3PO_4 + 3CaSO_4$

10 a) $HCl + KOH \longrightarrow KCl + H_2O$

b) Chloride ions and potassium ions

c) $H^+ + OH^- \longrightarrow H_2O$

11 a) $KOH + HNO_3 \longrightarrow KNO_3 + H_2O$

b) $NH_3 + HCl \longrightarrow NH_4Cl$

c) $3NH_3 + H_3PO_4 \longrightarrow (NH_4)_3PO_4$

12 a) potassium hydroxide + sulfuric acid $\longrightarrow$
$$\text{potassium sulfate + water}$$
$$2KOH + H_2SO_4 \longrightarrow K_2SO_4 + 2H_2O$$

b) sodium hydroxide + nitric acid $\longrightarrow$
$$\text{sodium nitrate + water}$$
$$NaOH + HNO_3 \longrightarrow NaNO_3 + H_2O$$

c) potassium hydroxide + phosphoric acid $\longrightarrow$
$$\text{potassium phosphate + water}$$
$$3KOH + H_3PO_4 \longrightarrow K_3PO_4 + 3H_2O$$

d) ammonia + sulfuric acid $\longrightarrow$ ammonium sulfate
$$2NH_3 + H_2SO_4 \longrightarrow (NH_4)_2SO_4$$

End-of-chapter questions

1 a) 16.47/16.5%

b) 17.07/17.1%

c) 21.21/21.2%

d) 35.0%

2 52.34/52.3%

3 a) i) 60

ii) 46.66/46.7%

b) $2NH_3 + CO_2 \longrightarrow CO(NH_2)_2 + H_2O$

4 a) Ostwald process

b) Platinum

c) KNO_3

5 a) i) Hydrogen

ii) Haber process

b) At temperatures below 450 °C the reaction is too slow to be cost effective.

c) Ammonium nitrate

d) In countries with a high rainfall Nitram would be washed out of the soil before it had a chance to be absorbed by the crops.

6 a) Any alkali, e.g. soda lime, NaOH, KOH etc.

b) i) Platinum

ii) The reaction is exothermic so it provides its own heat.

7 Nitrogen: 21.21/21.2%
Phosphorus: 23.48/23.5%

8 a) i) 20%

ii) 26.49/26.5%

b) Calcium phosphate is insoluble in water so could not be absorbed by the crops. If calcium dihydrogen phosphate is used as a fertiliser, it must be soluble in water; it also contains a higher percentage of phosphorus.

Chapter 12 Nuclear chemistry

In-text questions

1 ^{39}K protons = 19, neutrons = 20; ^{238}U protons = 92, neutrons = 146

2 Uranium

3

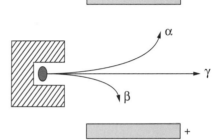

Alpha particles are positively charged so are attracted to the negative electrode. Beta particles are negatively charged so are attracted to the positive electrode. Gamma rays have no charge so are unaffected by the electric field.

4

Name	Alpha (α)	Beta (β)	Gamma (γ)
symbol	^4_2He	$^0_{-1}e$	γ
mass	4 mass units	0 mass units (approximately)	none
charge	positive (2+)	negative (1−)	none
penetrating power	low (sheet of paper)	higher (thin metal sheet)	very high (several cm of lead)

5 a) C

b) A

c) D

6 a) $a = 234$; $b = 90$, X is thorium/Th

b) $a = 32$; $b = 16$, X is sulfur/S

c) $a = 216$; $b = 84$, X is polonium/Po

7 Change 1 is alpha emission because the mass number has decreased by 4 units and the atomic number has decreased by 2 units.

Change 2 is beta emission because the mass number is unchanged and the atomic number has increased by 1 unit.

8 a) $^{3}_{1}H \longrightarrow ^{3}_{2}He + ^{0}_{-1}e$
   b) $^{224}_{88}Ra + ^{4}_{2}He \longrightarrow ^{228}_{90}Th$
   c) $^{60}_{27}Co \longrightarrow ^{60}_{27}Co + \gamma$
9 7.5 years
10 8 days
11 D

End-of-chapter questions

1 a) $^{14}_{6}C \longrightarrow ^{14}_{7}N + ^{0}_{-1}e$
   b) Fossil fuels are too old.
2 a) $^{35}_{16}S \longrightarrow ^{35}_{17}Cl + ^{0}_{-1}e$
   b) 174 days
3 a) $a = 234$, $b = 91$, X is protactinium
   b) $a = 206$, $b = 82$, X is lead
   c) $a = 214$, $b = 83$, X is bismuth
4 a) Cobalt-60 or iodine-131
   b) Americium-241 has a long half-life and is an alpha emitter so radiation is absorbed by the smoke detector's plastic cover.
   c) γ radiation has a high penetrating power and would pass through the paper too easily.
5 In the upper atmosphere, nitrogen atoms capture neutrons to produce the radioisotope carbon-14. The concentration of carbon-14 atoms in the atmosphere is fairly constant because, as they form, they combine with oxygen and the CO_2 produced is absorbed by green plants on the planet's surface. In living plants, the carbon-14 is slowly but continuously decaying and it is simultaneously being replaced by CO_2 absorbed during photosynthesis. When a plant dies, it stops taking in carbon dioxide, but the carbon-14 already in the plant keeps on decaying. If we assume that the ratio of carbon-14 to carbon-12 in trees today is the same as it was in the past, then by measuring the radioactivity due to carbon-14 in a sample of old wood and in the same type of new wood, scientists can estimate the age of the old wood.
6 a) Cosmic rays/radon/granite rock
   b) In the stars

Chapter 13 Chemical analysis

End-of-chapter questions

1 A is copper(II) sulfate; B is potassium chloride; C is sulfuric acid; D is silver(I) nitrate
2 a) $Cu(OH)_2$
   b) A fine white precipitate would form; the colour of the precipitate would be masked by the blue colour of the solution.
   c) The sample to be tested is mixed with a few drops of hydrochloric acid. A flame test rod is dipped into the mixture then held in the edge of a blue Bunsen flame and the colour of the flame is recorded.
3 a) A pH meter
   b) i) A pipette and a burette
      ii)

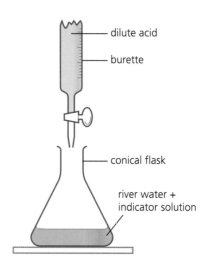

      iii) When the river water had been neutralised, the indicator would change colour.
   c) Accurate titre = (14.5 + 14.6)/2 = 14.55 cm^3
   d) River water is a mixture and there may be several substances in the mixture that react with the acid during the titration. It is not valid to assume that the only substance reacting with the acid is a substance that releases hydroxide ions.